C000225395

Properties and use of coal fly ash

Properties and use of coal fly ash

A valuable industrial by-product

Coal fly ash, or pulverised fuel ash, from coal-fired power stations: the production, properties and applications of the material

Compiled and edited by

Lindon K. A. Sear
United Kingdom Quality Ash Association

Published by Thomas Telford Publishing, Thomas Telford Ltd,
1 Heron Quay, London E14 4JD.
URL: http://www.thomastelford.com

Distributors for Thomas Telford books are:
USA: ASCE Press, 1801 Alexander Bell Drive, Reston, VA 20191-4400, USA
Japan: Maruzen Co. Ltd, Book Department, 3–10 Nihonbashi 2-chome, Chuo-ku, Tokyo 103
Australia: DA Books and Journals, 648 Whitehorse Road, Mitcham 3132, Victoria

First published 2001

A catalogue record for this book is available from the British Library

ISBN: 0 7277 3015 0

Produced by Gray Publishing, Tunbridge Wells, Kent
Printed and bound in Great Britain by MPG Books, Bodmin, Cornwall

Contents

Preface

Coal fly ash, or pulverised fuel ash (PFA) as it is known in the UK, from coal-fired power stations has been produced for many years and many research projects and papers have been published on the subject. It is estimated that some 300 papers are published per annum with the words 'fly ash' among the keywords. Therefore, it is surprising that there is no general textbook on the subject. We hope that this book will address that omission.

Coal fly ash is known by a variety of names including fly ash, pulverised fuel ash and as a coal combustion product (CCP). Within the UK, the accepted term and the most descriptive one is PFA. However, in general usage in many countries the term 'fly ash' is used for pulverised coal ash but it can also cover ashes from burning other materials. Such 'fly ash' derived from the combustion of other materials may have significantly differing properties and may not offer the same advantages as ash from burning pulverised coal. It is also necessary to define the type of coal used in the power station. Throughout this text, only fly ash from bituminous or hard coals is considered. Such coals produce siliceous fly ash, that is with less than 10% total calcium oxide (CaO), or class F fly ash using the USA definition. Calcareous fly ashes, that is those with more than 10% CaO or class C fly ash, may have significantly differing properties.

Throughout this text, when referring to coal fly ash, the preferred term 'fly ash' is used. However, the exception to this rule is when referring to a particular standard that incorporates the term 'PFA' within its title, e.g. BS 3892: 'Pulverised Fuel Ash as a Type II addition'.

The use of fly ash is not new. Recently published research[1] considered the properties of lightweight concrete made using fly ash that was developed 2000 years ago in the ancient culture of Totonacas near the modern city of Veracruz, Mexico. At approximately the same time, but independently, the Greek and Roman civilisations developed lightweight pozzolanic 'cement', in the case of Rome founded on the use of pozzolana from the village of Pozzouli, near Naples. The basic properties of pozzolanic materials, whether volcanic ash or power station fly ash, are still the same. The material can be used to produce more durable concrete, as a structural fill material, for landscaping sites scarred by industrial development and for the manufacturing of building blocks, precast concrete elements, etc.

The production of fly ash, its physical and chemical properties, various applications and environmental impacts are all considered in some depth.

Reference

1. Rivera-Villarreal R, Cabrera JG. The microstructure of two-thousand year old lightweight concrete. *International Conference,* Gramado, Brazil, 1999.

Acknowledgements

Compiled and edited by

Lindon KA Sear, BSc (Hons), PhD, FICT. Technical officer of the United Kingdom Quality Ash Association, a trade association representing the UK coal-fired power stations and users of coal fly ash.

Contributors

Mr Robert Coombs, BSc (Hons), CChem, MRSC. Head of the National Ash Laboratory. For contributing the chapter on fill and much of the information on the environment, and for proof reading the text.

Dr Guy R Woolley, PhD, CEng, FICE. Chartered civil engineer. For contributions to a number of chapters and proof reading plus his advice on format.

Mr Ken Swainson. Head of National Ash. For proof reading and contribution of information.

Mr Allan Foster, MICT. Technical manager for Rugby Cement and Ash Resources. For contributing towards the lightweight aggregates section.

Mr Steven Rule BSc, CEng. Civil engineer for P Forker Construction Ltd. For contributing towards the section on grouting.

Dr Robert A Carroll, PhD, CChem, MRSC. Technical manager for Marley Building Materials. For contributing the section on aerated concrete blocks.

Dr Trevor Grounds, PhD. Research and development manager for Tarmac Topblock Ltd. For contributing the section on furnace bottom ash in block manufacture.

Mr Andrew Weatherley, BSc, CEng, MICE. Civil engineer for Powergen, Power Technology. For contributing a section on the environmental aspects of fly ash.

Mr Chris Bennett, AMICT, MCS. Technical sales representative for ScotAsh Ltd. For proof reading various sections of the book.

Mrs Janice M Sear, BSc. For proof reading the text several times and correcting numerous grammatical problems.

Chapter 1

The production and properties of fly ash

Introduction

Fly ash has been used for many years for a wide range of construction applications. Its uses range from a cementitious material in concrete to a simple fill material. As a by-product material it has been reviled, abused, researched, researched again, praised and criticised, and yet it is often the ideal material for many applications. Fly ash for the purposes of this book is defined as follows:

- Fly ash is the ash resulting from the burning of pulverised bituminous, hard coals in power station furnaces.
- The furnaces are used to generate steam for the production of electricity.
- The furnace temperature is typically >1400°C.
- The resulting material is a siliceous ash consisting of the oxides of silica, aluminium and iron, and containing <10% calcium oxide. Many countries categorise siliceous fly ash as class F.

Coal is a readily available source of energy consisting of carbon and a mixture of various minerals (shales, clays, sulfides and carbonates). Coal, a mineral substance of fossil origin, may be one of four main types:

- anthracite (>90% carbon)
- bituminous or hard coal (~80% carbon)
- lignite and brown coal (<70% carbon).

With one exception (anthracite), only bituminous or hard coal is burnt in UK power stations. Hard coal originated in the Carboniferous period, part of the Palaeozoic era, about 345 to 280 million years ago, from vegetable matter (trees and ferns) which has been compacted and heated by geological processes.

During the mining and subsequent processing, washing reduces the siliceous material derived from the soil and minerals in which the trees were growing. Most of the material from which ash is formed is extraneous material, such as shale, but some is inherent in the coal and cannot be removed. Even after this processing the coal as delivered to power stations still contains about 15% of ash by weight after combustion.

Available reserves of coal in the world are estimated at more than 400 years' supply. In the UK, coal was traditionally obtained from deep mines. However, during the 1990s such coal became increasingly expensive to obtain relative to open-cast mined coals. In many countries high-quality coal exists near the surface and simply requires the overburden's removal and extraction. These coals are obtained from Australia, South Africa, South America and Eastern Europe at relatively low cost and widely exported throughout the world. Although coal is a readily available source of energy, it is predominantly carbon and, when burnt, produces carbon dioxide (CO_2). In order to reduce the effects of global warming many countries agreed to reduce CO_2 emissions in the Kyoto agreement of 1992. This has led to a move towards fuels richer in hydrogen and nuclear-based fuels, e.g. natural gas and nuclear generation. Owing to these procurement changes within the UK, overall CO_2 emissions were reduced[1] by 19% between 1970 and 1997, with 48% reductions being attributed to industry and 26% to domestic consumption. However, these environmental gains are offset by a large increase in emissions from UK transport, with an 87% increase occurring during the same period. At the time of writing coal-fired power generation still represents some 30% of UK electricity production.

Within the UK, coal-fired power generation was rapidly expanded after World War II to satisfy the needs of an increasing population with higher aspirations. Electricity output reached its peak in the early to mid 1970s with sufficient coal-fired electricity capacity to produce some ~16,000,000 tonnes of fly ash per annum. In the early 1970s, the UK deep mining industry went through a period of rationalisation. Over the next 15 years there was a decline in the UK deep mining industry until there was only a handful of deep mines left at the time of writing. Continued economic coal-fired generation will inevitably involve increasing levels of imported coal. This may have some effect on the resulting ash, but since all power stations in the UK are designed to handle UK coals, any imported coal will have to have similar physical and chemical properties, thus limiting the effect on the properties of the fly ash.

The pattern of coal-fired generation has changed over recent years (Fig. 1.1), in order to reduce CO_2 and other 'greenhouse' gas emissions. The introduction of gas-fired generation meant that many coal-fired stations ceased to be base load stations, that is those that are continuously operated, but adopted the double-shift system.

To minimise the cost of keeping major coal-generated plant running at full power, boilers are closed back or banked when not required for generation. Effectively, the boiler remains fired but at a lower level and virtually unable to produce the steam necessary to drive the turbines. When called on to generate, depending on consumer demand, the boiler will be brought

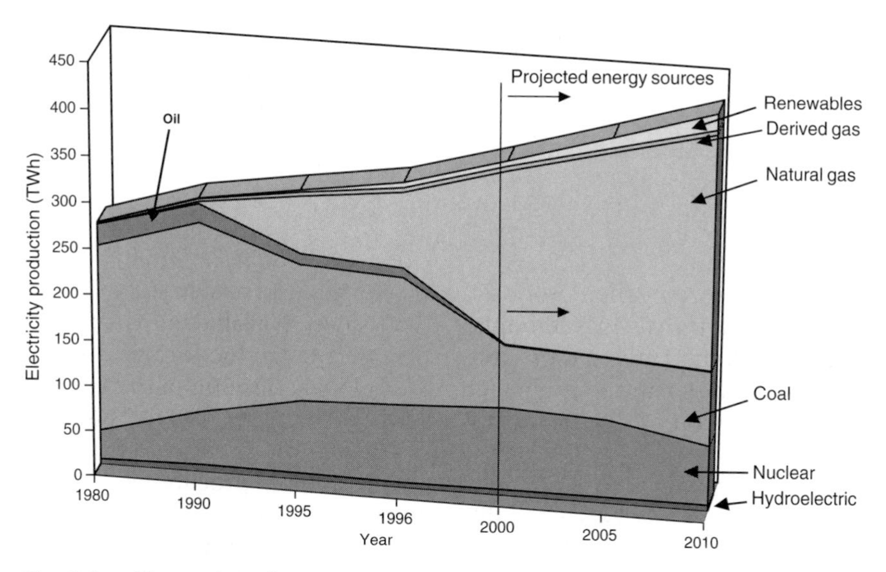

Fig. 1.1. Change in primary energy sources: electricity production from coal has reduced in recent years

back to full load. However, in normal circumstances it takes a relatively long time to raise steam and pressure. To reduce this time, oil is injected into the furnace, thereby boosting furnace temperature and making it more quickly available for normal coal injection.

To a small degree, the quality of fly ash produced is affected. During this 'boosting' period, a small increase in unburnt carbon or loss on ignition (LOI) has been found. In normal circumstances, the fly ash in collecting bunkers will intermix, but where it is taken directly to a site silo, consumers should be aware that there may be a small rise in LOI. It is therefore considered prudent to acknowledge this feature and check the LOI of fly ash produced during the run-up period of the boilers.

During the mid-1990s the electricity pool system was adopted within the UK. This operates by different generating companies offering electricity to the national grid at prices they think appropriate. The cheapest supplier is then contracted to supply for a period of time. Consequently, base load power stations are predominantly natural gas or nuclear. The advantages of gas-fired generation include:

- high efficiency: the gas stream and the heat generated can be used to extract energy
- reduced emissions: natural gas is low in carbon and high in hydrogen; CO_2 emissions are considerably less than from coal
- low capital costs: they are physically smaller, simpler and more compact than the equivalent coal-fired stations.

Nuclear generation supplies the base load because

- the nature of nuclear fission is such that switching off a reactor is effectively a waste of fuel
- the high capital and decommissioning costs need to be minimised by spreading over a high output
- the unit price of the electricity is low as a result of the way such generation was financed by the government through the 'nuclear levy'.

Coal-fired power stations satisfy the peaks in electricity demand, e.g. at breakfast and in the early evening. As many stations operate a number of generation sets, it is common for one set to be run to operate continuously, with the others being started when required by electricity demand.

Other environmental and economic aspects of coal-fired generation are emissions of sulfur dioxide and nitrogen oxides (NO_x). The emissions of NO_x have been reduced by installing low NO_x burners that reduce the temperature of the flame by staging the addition of the air to the coal. The burners work by creating a two- or three-stage combustion process. A lower temperature fuel-rich region is created in which any NO_x formed is partially reduced back by nitrogen. Controlled amounts of excess air are introduced at a later stage to complete the burnout of the semicombusted coal particles. The overall effect on ash is generally seen as a slightly increased carbon content, because of the lower flame temperatures. Research studies carried out in Germany into the use of ash from low NO_x boilers found the ash to have no adverse effects on properties relevant to concrete technology.[2]

To reduce sulfur emissions more complex equipment involving reacting the acidic waste gases with calcium carbonate ($CaCO_3$) or quick lime (CaO) have been adopted. In the current UK coal-burning power stations two specialist calcium carbonate flue gas desulfurisation (FGD) plants have been installed. These plants use limestone as the absorbent, in an arrangement whereby flue gases are directed through a curtain of limestone slurry en route to the atmosphere. Sulfur dioxide in the flue gases combines with the high-quality limestone to create calcium sulfite and CO_2. This material is then passed through an oxygenation tower where calcium sulfate (gypsum) is formed.

The 'wet limestone' FGD plant is located in the flue gas emission route, positioned after the electrostatic precipitators but before entry to the power-station chimney. Because this secondary method of reducing sulfate emissions to the atmosphere is placed after fly ash has been extracted from the flue gases, installation of this type of desulfurisation plant has no effect on the quality of fly ash.

Desulfurisation equipment is large and expensive and adds the problems of consuming large quantities of carbonate aggregates. In addition, a market for the resulting gypsum is required. Because of these changes, the relative economic efficiency of coal-fired generation is being degraded in favour of gas and nuclear generation. However, supplies of natural gas are typically quoted as 20 years and nuclear generation carries the environmental baggage of disposal, safety, etc. Whether this difference in perception of coal-fired generation will continue depends on a large number of factors. Several questions will have to be answered: will more nuclear generation be allowed; will energy conservation begin to have significant effects on electricity usage; will new natural gas supplies be found, etc.?

Coal-fired electricity generation

A coal-fired power station (Fig. 1.2) is used to generate heat, steam and electricity. The typical schematic layout for a power station is shown in Fig. 1.3. Steam is raised in modern power stations using coal which, before combustion, is ground (i.e. pulverised) in mills of various types. Coal, a mineral substance of fossil origin, may be one of four main types: anthracite (>90% carbon), bituminous, lignite and brown coal (70% carbon). With the exception of anthracite, only bituminous coal is burnt in UK power stations. It consists of carbonaceous matter and a mixture of various minerals (shales, clays, sulfides and carbonates).

Coal is delivered to power stations as 'smalls', i.e. in lumps of about 50 mm diameter or less. It is stored in heaps where it is compacted to

Fig. 1.2. Typical UK coal-fired power station (courtesy of English Partnerships)

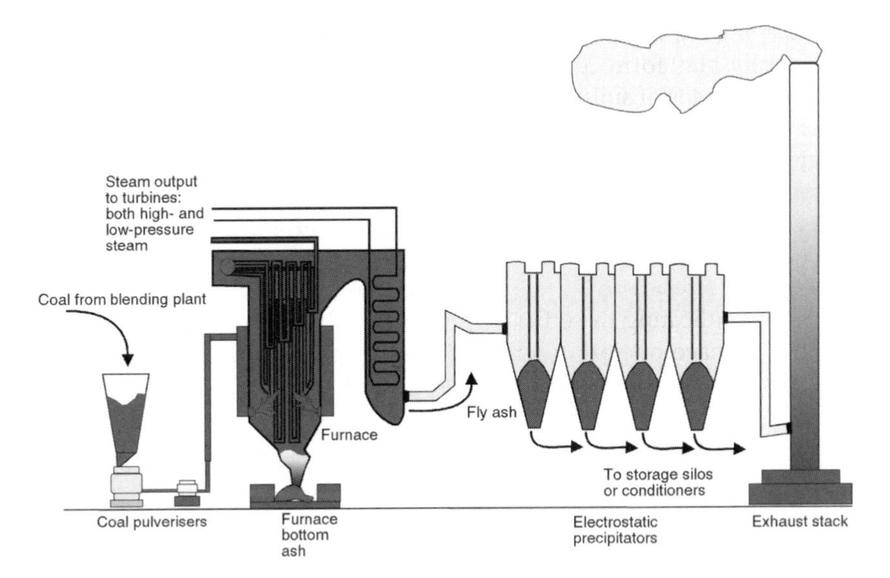

Fig. 1.3. Schematic layout of a power station

Table 1.1. Reduction in coal particle size after pulverising

Sieve size	Before pulverising (% by weight passing)	After pulverising (% by weight passing)
12 mm	94·3	
6·4 mm	80·9	
3·2 mm	64·4	
1·6 mm	39·5	
790 μm	18·8	
150 μm	–	96·0
100 μm	–	92·0
75 μm	–	80·0

prevent any spontaneous combustion or oxidation. The coal is recovered and placed in bunkers that feed the power station's coal mills. These grind the coal to a size 70% passing 75 μm. This finely ground coal is transported in a current of heated air (primary air) to the burners, where it is blown into the boiler. The degree of size reduction can be seen in Table 1.1.

The pulverised coal is injected into the furnace in a stream of hot air, as shown in Fig. 1.3. This air is heated by a heat exchanger using the furnace exhaust gases. The coal burns in a multistage process suspended in the combustion air in the boiler, reaching a peak temperature of some 1450 ± 200°C. This temperature is above the melting point of most of the minerals present, which undergo various chemical and physical changes. For example, clay forms glass spheres of complex silicates; pyrites is converted into oxides of sulfur and iron, including spherical particles of magnetite; and aluminium oxidises. The molten mineral matter forms into spheres, which rapidly cool to below melting point and are frozen in an amorphous glass (Fig. 1.4). Around 1–2% by weight of the fly ash occurs as cenospheres that consist of silicate glass in which the silica content is higher but the calcium content lower than fly ash.

The average residence time for a particle of coal within the furnace is only 3–4 s,[3] indicating the efficiency of this method of burning pulverised coal. The furnace operating temperature compares with ash fusion temperatures measured in the laboratory of about 1150°C for initial softening to just over 1550°C for fully molten. The ash produced from the coal combustion is, therefore, molten at the end of combustion and is still in suspension in the furnace gases. The fly ash continues to be transported by the combustion gases (now the 'flue gas') through the convection parts of the boiler and is captured in an electrostatic precipitator at the boiler outlet.

The exact nature of the fly ash depends on a variety of factors including the temperature, the type and fineness of the coal, and the length of time the minerals are retained in the furnace. Approximately 80–85% of the ash carried out of the furnace by the exhaust gases is subsequently extracted by mechanical and electrostatic precipitators. The remaining 15–20%

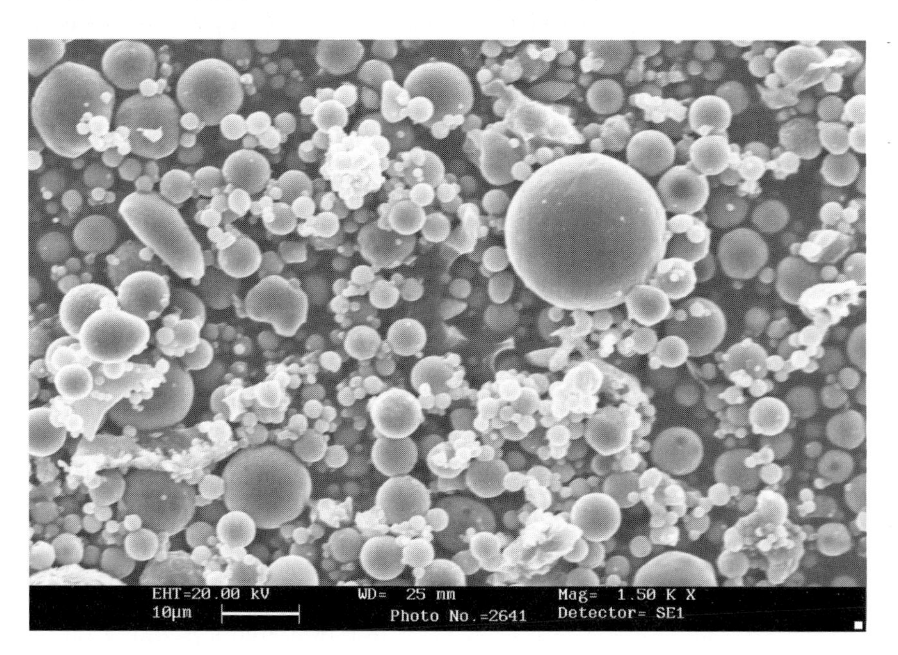

Fig. 1.4. Fly ash particles are spherical

condenses on the boiler tubes and subsequently falls to the bottom of the furnace where it sinters to form furnace bottom ash (FBA).

In the UK, FBA is flushed from the bottom of the furnace using water. It passes through a crusher and is then delivered to ash pits to drain. This material is then loaded directly into tippers where it is predominantly used for the manufacture of concrete building blocks.

The gases from the furnace may pass through cyclone filters or mechanical filtration systems which remove the coarser and heavier fractions. All gases pass through electrostatic precipitators that are connected in series to remove the finer and lighter materials. Here the flue gases, now at a temperature of about 130°C, pass between plates at earth potential and electrically charged wires at a potential of between 40 and 50 kV negative to earth. The resulting corona current is carried by ions, which collide with the ash particles making them negatively charged and attracted to the earthed plates. A total current of 200–400 mA results. Nowadays, the specification for an electrostatic precipitator is based on outlet dust burdens set by legislation (e.g. $<50\,mg/m^3$). However, in the past, the collection efficiency was specified and this was usually greater than 99·3%, which gives a better impression of the effectiveness of the plant. The ash is removed from the plate by 'rapping', which is by blows from a series of mechanical hammers that hit the plates at defined intervals. The ash falls into hoppers at the bottom of the precipitator and is then removed dry.

Some unburnt or partially burnt carbon residue is collected with the fly ash in the precipitators. The amount depends on the nature of the coal, its fineness after being pulverised, and the design and configuration of the furnace and precipitators. Many UK furnaces designed before 1975 were intended to extract the maximum energy from the coal. Over more recent years increasingly stringent environmental requirements have required slightly lower furnace temperatures and the progressive injection of air to reduce the degree of nitrogen oxidation, the so-called low NO_x burner. The drawback of such burners can be a higher LOI[4] in the resulting fly ash. LOI is a measure of the unburnt carbon, which remains in flue gases. This increase in LOI is not simply due to the retrofitting of low NO_x burners to older furnaces. It is clear from reports throughout the world that power companies are experiencing similar problems[5] even with newer stations, with a doubling of the LOI being found in some cases.[6]

To improve further the efficiency of electrostatic precipitators ammonium and sulfur compounds may be injected into the exhaust gas stream before the precipitators to encourage ionisation and agglomeration of the ash. Sulfur injection improves the surface charge on the alumino-silicate material. It is for this reason low sulfur coals can reduce SO_2 emissions but tend to be prone to increased particulate emissions. Ammonia injection encourages agglomeration for particles that do not hold a charge. Consequently, the precipitators remove the ash more efficiently, reducing the particulates in the flue gases. Ammonia injection can have a detrimental effect on the fly ash, with a slightly increased sulfate content and a distinctive ammonia smell. The ammonia is released when the fly ash is exposed to water, especially in an alkaline environment as in concrete. Although the levels of ammonium injection are invariably very low and there are no discernible deleterious effects on performance, e.g. in concrete, at these levels the problem with the odour can be significant.

At some stations, the earlier electrostatic precipitators in the series tend to extract a larger fraction of the coarser fly ash and more carbon. Some stations are able to keep the various fractions of fly ash separate whereas others are forced to combine the ash into a single output.

FGD has been installed at some power stations. In the limestone/gypsum process, the flue gases, on exiting the precipitators, are passed through a sprayed slurry of limestone to remove the sulfur dioxide before passing into the chimney. In the UK, the resulting gypsum is used predominantly for the manufacture of plasterboards for general construction purposes. An alternative system using a spray-dried system uses lime, with the resulting gypsum being collected in bag filters.

After extraction from the flue gases fly ash may be treated in a variety of ways as follows:

- By being pumped using air into storage silos as a dry powder. This material may then be used in concrete with or without selection (see section on EN 450 fly ash and BS 3892 Part 2, Pulverised fuel ash in Chapter 3) or may be classified to make a finer product (see section on BS 3892 Part 1, Pulverised fuel ash in Chapter 3).
- By being 'conditioned': a small quantity of water is added to produce a dampened material, which can be handled and transported without problems with dust. The amount of water can be adjusted to suit the end use.
- By slurrying in copious quantities of water where it is pumped to lagoons. Here the ash is allowed to settle. These lagoons can sometimes be drained off and the ash recovered either for use or for disposal. Lagoon ash may contain some FBA.

The various methods of extracting the fly ash from the furnace gases result in fly ashes of differing particle size distributions.

Processing fly ash for use in concrete

Processing fly ash by classification is designed to optimise the characteristics for use in concrete as per BS 3892[7] Part 1 'Specification for pulverised-fuel ash for use with Portland cement'. The pozzolanic reaction (described in Chapter 3) of fly ash with lime depends on the surface area exposed to the lime-saturated pore solution within the concrete. In addition, the spherical nature of fly ash and the packing/optimisation effect of classified fly ash particles reduce the water requirement for a given workability. In general, by selecting the ash or by removing the coarsest fractions from within 'run of station' fly ashes three benefits are produced:

- increase in surface area per unit mass, resulting in increased pozzolanic activity and thus improving the rate of strength gain
- reduction in the water demand of the concrete for a given workability
- reduced variation in the concrete by improving the consistency of the particle size distribution.

Both selection and classification were used to produce BS 3892 Part 1 PFA. Selection involves monitoring the fineness and LOI of the fly ash and diverting the ash to a differing storage silo when of suitable fineness. However, in recent years selection techniques have not been used because they proved to be unreliable at times. Classification currently predominates. With the advent of EN 450[8] and to prevent any conflict with BS 3892 Part 1, classifying fly ash became mandatory for compliance with BS 3892 Part 1. Figure 1.5 shows the typical design of an air-swept classifier.

Typically, a 'run of station' fly ash, that is with no processing, will have 25% retained on the 45 μm sieve and the fly ash post-classification will have 8% retained on the 45 μm sieve. Some 17% of the product will be removed and sold as a fine aggregate or fill material and the balance is

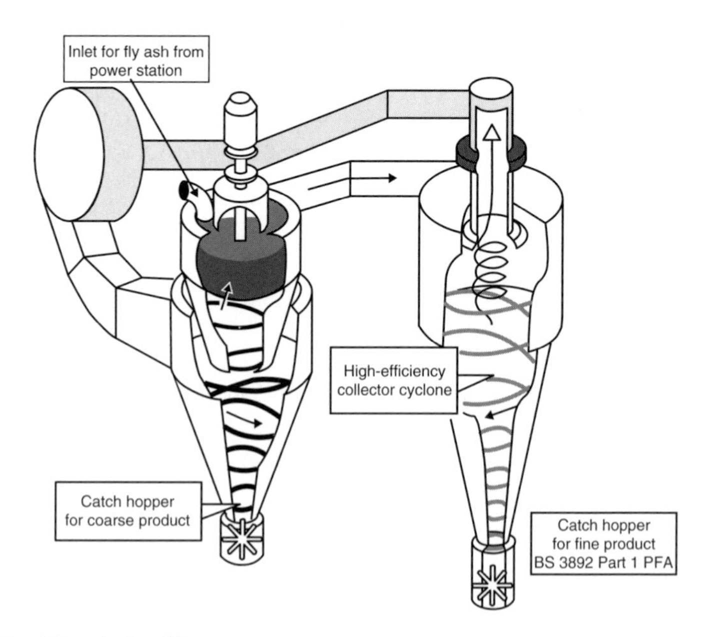

Fig. 1.5. Fly ash classifier system

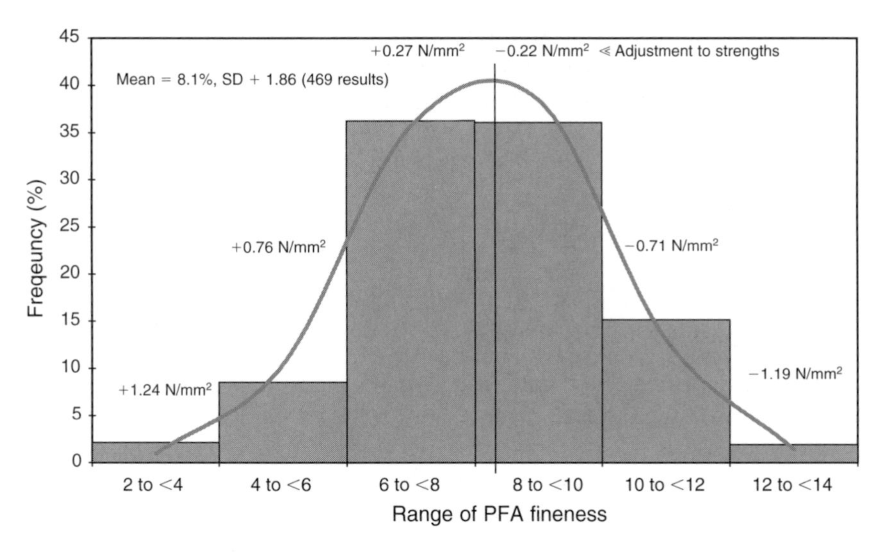

Fig. 1.6. Strength and fineness distribution of UK BS 3892 Part 1 fly ashes in 1997 (from mortar prism test data; see BS 3892 Part 1)

classified fly ash to BS 3892 Part 1. The reject product from the classifier, called the grits, may be ground to increase the fineness, enabling it to be sold as fly ash to BS 3892 Part 1.

It must be realised that the use of the percentage retained on the 45 μm sieve only gives an indication of the particle size distribution or the surface area of a fly ash. Therefore, it does not follow that after classification of fly ash all sources will give equal reactivity. Fortunately, within the UK the fly ashes used in concrete are all siliceous (class F) and produced from similar designs of power station, using similar hard coals. Consequently, UK classified fly ashes have a remarkably consistent strength for similar fineness values, as indicated in Fig. 1.6.

Conditioned fly ash

For ease of handling fly ash can be 'conditioned' by adding a controlled amount of water to prevent dust blow problems. Conditioners can consist of screw-type continuous mixers that rely on a constant supply of water. Batch-type mixers are also used which give a better control of the final moisture content. This is important where fly ash is being used for fill applications where the moisture content is critical in achieving maximum compaction. However, as the fly ash is normally still warm some moisture may be lost through evaporation, and water spraying on site may be required to obtain this optimum moisture content.

Disposal schemes using conditioned ash are designed for the transport of fly ash from the power station by either road vehicle or conveyor in conditioned form. Moisture content, usually between 8% and 14%, is based on the need to prevent dust blow. This is also to ensure that it can be adequately compacted at the disposal site. Mist spraying of water or fixation of the surface material using chemicals may be required on such stockpiles to prevent further dust-blow problems until the material is removed or capped.

If the disposal scheme involves the filling of worked out mineral deposits, topsoil and subsoil will normally have been stripped and stored next to the void before mineral extraction. Should disposal be taking place above ground to create a mound on virgin ground, soils should be stripped and stored prior to filling and placed for eventual use in final restoration works.

Once site preparation works have been completed, which for above-ground disposal may include a drainage blanket of granular material and carrier drains, deposition of fly ash may commence. Fly ash is placed and compacted using conventional earth-moving equipment. The thickness of succeeding layers of PFA should be determined early in the disposal programme. The fly ash is profiled to final levels and, upon completion of filling to these levels, the topsoil and subsoil previously stored can be recovered and spread over the ash to a minimum depth of 0·3 m. Cultivation of the area may then begin.

A prime example of this type of above-ground scheme is found at the Drax Power Station, where surplus fly ash is deposited at the Barlow Ash Mound Landscape Reclamation Scheme. As final levels are completed topsoil is spread and grass and shrub/tree planting follows. The scheme, begun in 1974, is designed to produce over 160 ha of cultivated landscaped mound when completed.[9]

It is not normal for earth embankments, or bunds as they are known, to be constructed around a disposal area. They may be required, however, if the disposal site lies within the wash lands of a river to prevent fly ash being washed out by floodwaters. Suitable drainage channels and settling pond facilities will be required around surface mounds created by disposal of fly ash.

Lagoon fly ash

An alternative to stockpiling conditioned fly ash is to slurry it with water and pump it to lagoons, where it is allowed to settle. Normally, ratios between 10 : 1 and 3·5 : 1 water to fly ash are used. The more modern plants use less water. Lagoons are of two types: those large enough to hold fly ash until sale and those small enough to require regular emptying and stockpiling of the reclaimed material. Discharge into a lagoon causes segregation, with the coarser material concentrating at the inlet and the finer at the water outlet. The pattern of particle sizes can become highly complex if there are multiple inlets or outlets.

At the disposal site it is necessary to construct bunds in order to contain the slurry. If the scheme involves the filling of worked out mineral deposits, the bunds will normally be constructed using soils and subsoils that were stripped prior to mineral extraction taking place. Alternatively, if the scheme involves the disposal of fly ash above ground level, soil and subsoil will be stripped from the disposal area and suitable material will be imported to raise the bunds to their required height. Conditioned ash may be suitable for this purpose, a method successfully used at the Gale Common site in North Yorkshire, which has taken ash from the Ferrybridge and Eggborough power stations.[9]

When pumped to a lagoon the fly ash particles will settle out while the supernatant water will be decanted into a buffer pit or settling lagoon before being discharged to a watercourse. In this way, suspended solids should settle out and sufficient dilution will have taken place to ensure that the discharge to the watercourse meets the relevant environmental standards. When the fly ash has reached its final level pumping will cease and the fly ash will be allowed to dry out. To prevent dust blow the surface should be served with a dust suppressant. When the fly ash has dried out sufficiently the soils contained in the surrounding bunds will be spread over the surface of the ash. Cultivation of this soil may then begin and the area be returned to agriculture.

A characteristic of lagoon deposited fly ash concerns the variable nature in particle size of the deposit. As slurried ash is delivered into the lagoon from the pipeline, it will enter a void filled with a solution of around 70% water. Natural sedimentation dictates that the coarser particles will settle ahead of the finer particles. Consequently, the deposited ash will have a variable particle size profile consistent with the position of delivery and amount of solids in suspension at the time. Subsequent recovery of lagoon deposited fly ash, depending on utilisation, may require remixing the ash to achieve a more acceptable particle size gradient.

It should be emphasised that planning and environmental legislation must be strictly followed during the disposal of fly ash in this way. It should also be noted that where the capacity of a lagoon exceeds 5 million gallons (22·7 million litres), the generating companies in the UK have considered such a lagoon constructed above surrounding ground level as a reservoir. As such these fall within The Reservoirs Act, 1985, and this legislation must be followed.

The material in a lagoon will drain and approach optimum moisture content after a period of storage, as in Fig. 1.7. The speed of this process

Fig. 1.7. Fly ash in a drained lagoon

is influenced by various factors such as the ratio of transporting water to fly ash, the design of the outfall, groundwater levels and the method of operation. Some lagoons are kept flooded until full. This will prolong the drainage time when the ash is recovered. At some stations, on occasions, small quantities of FBA will be added to the fly ash/water mixture to scour the pipe work. To aid drying lagoon fly ash will be dug out while still very wet and piled up to encourage drainage and air drying. This technique is known as harvesting. The main uses of lagoon fly ash are in-fill and grouting, although both conditioned and lagoon can be used in concrete.[10]

Stockpiled fly ash

The daily make of fly ash from a power station may not be sufficient to meet the requirements of major fill contracts. To overcome this problem fly ash is stockpiled as in Fig. 1.8, either with ash directly taken from the conditioners or by removal from lagoons. In some areas, county and local authorities are prepared to project their usage of fly ash for fill purposes, e.g. for road construction projects. This allows for some forward planning

Fig. 1.8. Fly ash on stockpiles

of power station activities to meet this demand. During periods when production exceeds sales the fly ash is stockpiled, normally on existing tipping grounds. Material is tipped in such a way as to encourage some secondary mixing, which gives a more consistent material. This is particularly important with lagoon ash because of the variable nature of the particle size distribution within the lagoon. Lagoon and stockpiled material may be mixed to obtain the optimum moisture content required for fill purposes.

Stockpiled ash may harden owing to the presence of a small quantity of CaO (quicklime) in the ash. This hydrates with water, producing some heat and $Ca(OH)_2$ (lime). In the presence of CO_2, this carbonates to form $CaCO_3$ which acts as a binder. However, the lime content of UK ashes is generally low and unlikely to lead to much heat evolution when the ash is mixed with water. In addition to carbonation, the pozzolanic effect will cause hardening. If there is sufficient lime available and the ash is compacted so that CO_2 is excluded, the lime cannot carbonate. Then, some agglomeration of finer particles due to a pozzolanic reaction may occur, forming lumps of material. To reduce this hardening of the fly ash during the stockpiling operation it is necessary to tip over a high face greater than 3·5 m in height. This breaks up any hardened ash and ensures sufficient compaction to preclude the CO_2 required for hardening to occur. When fly ash is supplied from stockpiles, it is often necessary to use some screening plant to remove lumps of hardened ash (Fig. 1.9).

Fig. 1.9. Screening fly ash to remove the coarser fraction at the stockpile

Longer-term stockpiled fly ash

Considerable quantities of fly ash have been deposited through the years either in former mineral workings below normal ground level or in above-ground mounds. These stockpiles represent a valuable resource and this fly ash retains significant chemical and physical properties.[11] When recovering fly ash from stockpiles for use in concrete or grout, this material will have to be screened to remove any agglomerate.

To prevent dust-blow problems conditioned ash mounds and drained lagoons are normally protected in some manner. Hydra seeding of the surface is a useful option, where a controlled mixture of bitumen and selected grass seeds is sprayed over the surface. This treatment will prod[illegible] grass sward capable of growing on to a usable pasture. For permanent restoration, fly ash surfaces are covered with a depth of topsoil. The topsoil surface is sown with seed to produce pasture. Where seeding is undertaken, the seed mixture will contain seeds that are resistant to boron and are known to grow in this situation.[12]

Cenospheres

Cenospheres are unique free-flowing powders composed of hard-shelled, hollow, minute spheres. Their main characteristics are:

- they are hollow spheres
- the particle sizes range from 50 to 200 μm in diameter
- they have ultra low densities
- they exhibit low water absorption
- they have good thermal and electrical resistance.

Table 1.2. Chemical composition of cenospheres

Element	Percentage
Silicon (as SiO_2)	55–61
Aluminium (as Al_2O_3)	27–33
Iron (as Fe_2O_3)	4·2–9·5
Calcium (as CaO)	0·2–0·6
Magnesium (as MgO)	1·0–2·1
Alkalis (as Na_2O, K_2O)	0·5–4·6
Carbon (LOI – normally)	0·01–2·0

Between 1% and 2% of the fly ash produced from the combustion of coal in UK power stations is formed as cenospheres. The source of the coal source greatly affects the quantity of cenospheres produced. Cenospheres consist of silicate glass in which the silica content is higher but the calcium content lower than that of fly ash. The amount of soluble matter is very much lower, averaging 0·2%. The range of chemical composition is shown in Table 1.2.

Demand for cenospheres is such that they are extracted from the lagoon and sold to specialist processors. They dry them, grade them to single sizes and then sell them to a variety of industries for use as a strong lightweight filler material. Figure 1.10 shows a novel, though discontinued, technique for removing cenospheres.

Cenospheres have low density and low thermal conductivity. The bulk density, around one-quarter of that of fly ash, varies from 250 to 350 kg/m^3 with an apparent density of individual particles in the range of 400–600 kg/m^3. The thermal conductivity of the lightly tamped material measured 0·10 W/mC, determined at a mean temperature of 50°C (cold face 10°C, hot face 90°C). Cenospheres have a shell thickness of about 10% of their radius. The mean diameter is 100 μm, the range of diameters being 5% (by weight) below 50 μm and 20% above 125 μm. They are at the coarser end of precipitator fly ash grading. The initial sintering temperature is 1200°C i.e. 100–200°C higher than fly ash. However, unlike fly ash, sintering is accompanied by shrinkage and the spheres collapse at temperatures above 1300°C.

Applications

The main application of cenospheres is as an inert filler. With a density lower than water (typically 0·7), cenospheres provide up to four times the

Fig. 1.10. Collecting cenospheres from a lagoon at Peterborough landfill scheme

bulking capacity of normal weight fillers. The microspherical shape dramatically improves the rheology of fillers, whether in wet or dry applications. Cenospheres are extremely stable. They do not absorb water and are resistant to most acids. As it is a refractory material, the cenosphere can resist high temperatures.

Cenospheres can be used in plastics, glass-reinforced plastics, lightweight panels, refractory tiles and almost anywhere that traditional fillers can be used. Because of their flexibility, they are used in many high-technology and traditional industries, including aerospace, hovercraft, carpet backing, window glazing putty, concrete repair materials, horticultural use, and off-shore oil and gas production industries. In the aerospace industry, cenospheres have been used to manufacture lightweight propeller blades. The cenospheres reduce the weight but also increase the strength.

Furnace bottom ash

FBA forms around 20–25% of the ash produced. FBA is a coarse material which may be as large as 75 mm in diameter. It falls to the bottom of the furnace into a hopper and is removed using high-pressure water jets along

Table 1.3. Typical furnace bottom ash particle size distributions

Sieve size	Typical % passing	
	Coarse range	Fine range
37·5 mm	98	100
20 mm	85	95
10 mm	61	90
5·0 mm	44	70
2·36 mm	37	55
1·18 mm	29	50
600 μm	21	40
300 μm	14	25
150 μm	7	14

sluiceways. Coarse material is crushed down to <25 mm before the material passes to storage pits. It is normally loaded on to tippers and, in the UK, transported to the block-making factories. Typically, FBA has the particle size distribution shown in Table 1.3. Because of its size and the lightweight nature of the material, it has proven ideal for the manufacture of lightweight concrete blocks.

The future for fly ash production

Co-firing

One environmental aspect of burning carbon-based materials such as coal is the production of CO_2. This is believed to contribute to global warming and many governments world-wide have made undertakings to reduce CO_2 emissions to 1992 levels. In addition, there have been moves to reduce the amount of waste which is landfilled and considerable effort has been put into using alternative fuels which may be co-fired with coal. Materials currently being considered are:

- sewage sludge
- waste paper
- wood pulps, chippings, etc.
- rendered animal products, e.g. bone, fats and meat
- by-product solid, gaseous and solvent materials derived from a range of chemical processes
- domestic waste.

These materials may have a wide range of calorific values and result in varying amounts of fly ash. For example, co-firing solvents with coal at up to 70% by mass of the input material may only produce some 1% of the resulting fly ash. Conversely, a large proportion of the resulting ash may originate from the input material, which could have a very significant effect on the properties of the resulting fly ash. The chemical make-up of the co-fired material will have consequences for the chemistry of the final material. However, owing to the relatively high temperatures involved in a power-station furnace (1400°C) these materials would tend to be held in an amorphous glassy material, unlike in waste incinerators, which normally operate at much lower temperatures, typically 850°C. Consequently, there are considerable environmental benefits to firing at higher temperatures.

Considerable other difficulties may be encountered depending on the design and age of the furnaces. UK furnaces were designed to operate with a narrow range of coals and produce the maximum energy efficiency from them. With the advent of reducing NO_x and SO_x emissions, this efficiency has been degraded. Co-firing can result in a further reduction in efficiency or problems with the boilers, e.g. fouling of boiler tubes, excessive corrosion within the furnace or adverse effects on coal pulveriser efficiency. Although these problems are not insurmountable, in many cases they may add to the running costs and not compare favourably with other fuels, e.g. natural gas.

In addition to the above, co-firing of potentially hazardous wastes requires extra control systems, including:

- increased environmental monitoring when dealing with biological waste, e.g. sewage sludge
- additional tests to prove the effects, if any, of co-firing on resulting fly ash, e.g. strength and durability of concrete mixes, leaching tests
- enhanced quality assurance systems to show satisfactory levels of control.

Extraction of other substances

Fly ash contains a wide range of materials and examination of analyses suggests that some of these substances could be profitably extracted.

Iron extraction is comparatively easy using magnetic processes. These are able to extract up to half of the iron present. Both wet and dry methods have been used and a two-stage separation has given a concentrate containing ~50% Fe, 1·5% CaO, 0·3% P_2O_5, and 0·18% S. This concentrate is not considered a high-grade equivalent to iron ore and would require more purification before attracting a sufficiently high price. However, the concentrate has an alternative use as a heavy medium for coal washing, which it performs very well, and has an advantage over magnetite in that it can all be recovered by magnetic means. It could also be competitive as a dense material for use in nuclear shielding mortars and concretes.

Aluminium can be extracted from fly ash. One method that has been described makes use of the sulfating stage with concentrated sulfuric acid, followed by the production of potash alum from which the aluminium present is removed by a primary amine. Another process, which was operated on the European continent during World War II, makes use of the high-temperature reaction between lime and shales, which produces a water-soluble calcium aluminate.

Titanium (~1%) and germanium (20–100 ppm) are present in fly ash. Their extraction has proven not to be economically viable to date.

Carbon reduction

The unburnt carbon content of fly ash has increased with the introduction of low NO_x technology. Carbon in fly ash has some unfortunate effects on air-entrained concrete in that the air bubble structure is destroyed and large quantities of air entrainment admixture are required to compensate. In addition, excessive carbon can float to the surface of concrete and cause staining on slabs, especially with very fluid concretes. For fill applications, a variable carbon content affects the optimum density and moisture content, which can cause problems. Consequently, carbon removal has many attractions:

- The extracted carbon could be used as low-grade activated carbon[13] for filtration applications.
- The carbon could be used to fuel the furnace, e.g. recycled.
- Greater consistency in colour and quality of concrete, especially air-entrained concrete, would be achieved.

Various techniques are used to remove carbon. Froth floatation has been used to float the carbon to the surface of a fly ash/water mixture. The resulting materials require drying to be used for many applications. Levy[14] suggests a bubbling fluidised bed of air in conjunction with an acoustic field to aid particle separation. Strong segregation patterns are formed in the fluidised ash, from which a high carbon layer is removed. Levy's experiments showed that this equipment is capable of reducing carbon content by up to 50%.

Electrostatic extraction can be used where the fly ash is charged with a controlled electropotential. The fly ash is exposed to some form of positive charge and then allowed to free fall between charged plates. Stencel *et al.*[15] suggest that this charge can be obtained by contact rubbing, whereas other techniques charge the fly ash electrically. The carbon fraction gains a differing charge to that of the fly ash and can be separated from the mass as in Fig. 1.11. These techniques have been used commercially[16] to reduce carbon content to relatively low levels. Typically, LOI values will be ~2% even when the original fly ash has carbon contents in excess of

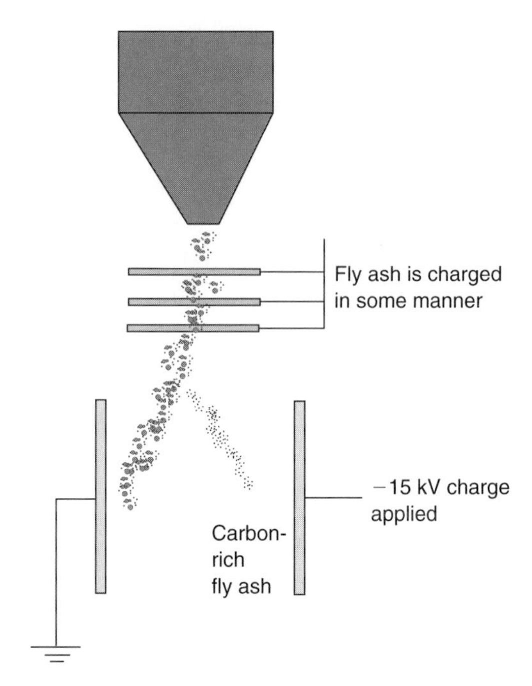

Fig. 1.11. Principles of electrostatic carbon removal

10%. Unfortunately, such techniques are only just commercially viable in the UK.

Carbon burn-out is achieving some acceptance in the USA.[17] The burn-out process used is a fluid bed combustion technology that reduces the carbon content of the fly ash to a desirable level and simultaneously returns heat energy to the power plant boiler for the production of steam to generate electricity.

High-performance fly ash

Ultrafine pozzolanas, such as silica fume, are used to produce high-performance concretes. The particle size of silica fume varies between 0·1 and 1 μm and dosage rates usually range between 5% and 15% by weight of cement. Concrete strengths in the range of 80–150 MPa are commercially feasible using silica fume. Fly ash from power stations has a small proportion of such fine particles, typically <10%, which may be commercially viable to extract. Micronisation of fly ash can be achieved either by air-swept classification or by grinding. The latter is the more expensive process but uses more of the fly ash. Vapourisation[18] at 2400°C and condensation of the fly ash is another possible method of creating such superfine fly ashes. Cornelissen[19] found that ground fly ash has no significant effect on the fluidity and produced concretes of high strength and density. Performance was similar to concrete made with silica fume at the same water/cement ratios.

Utilisation of fly ash

Fly ash is utilised in many ways in the UK (Fig. 1.12). World-wide there is a wide variance in utilisation, ranging from virtually all the ash being dumped to total usage. For example, in The Netherlands it is illegal to dispose of fly ash in a disposal site.

In addition to fresh production it is estimated that some 250,000,000 tonnes of fly ash exists in stockpiles throughout the country that could be used as alternatives to naturally occurring aggregates. The UK utilisation rate has remained stable at ~50% of production for a number of years.

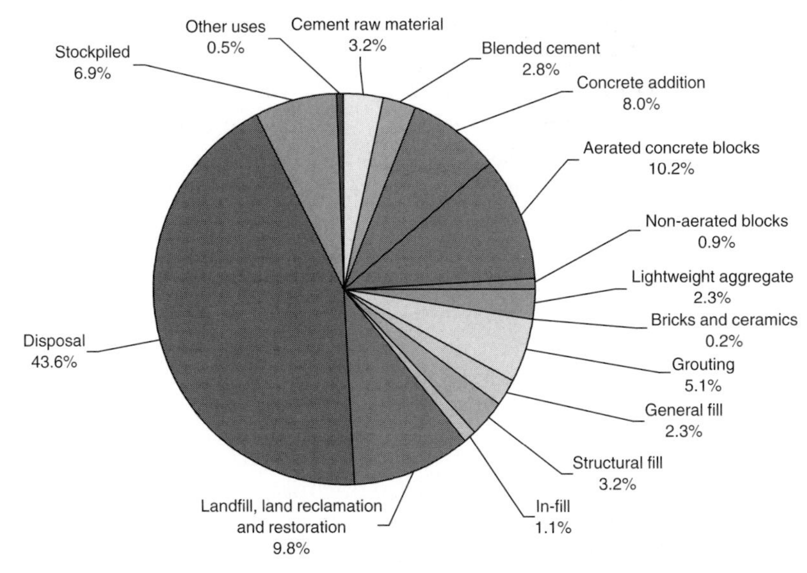

Fig. 1.12. Utilisation of fly ash from power stations (1997); approximately 50% of the fly ash produced in the UK is used

Particle size distribution and shape

Because of the way in which fly ash is produced, the particles, particularly those below 50 μm, are spherical in shape. As the coal is burnt producing temperatures in the region of 1400°C, the minerals associated with it become molten and form a spherical shape. Because of the rapid cooling

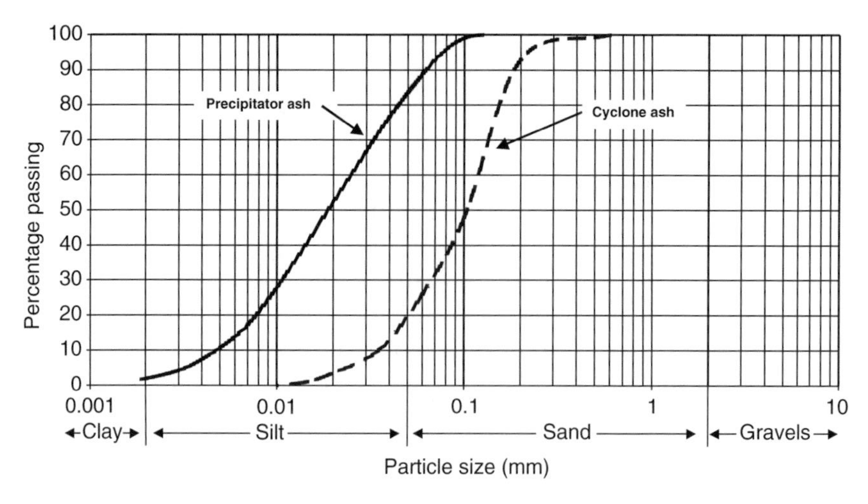

Fig. 1.13. Typical particle size distribution for UK fly ash (materials classification based on BS 1377 Part 1[20])

experienced by the fine ash particles as they pass out of the furnace, they solidify as an amorphous, glassy material in this shape.

Particles in the coarse silt/fine sand sizes have the potential to become airborne in certain conditions. Fresh conditioned and stockpile fly ash is like a fine-grained soil, and it is mainly silt-sized and generally acts like silt (Fig. 1.13). Finer fly ash has a silky feel, although a coarser one may feel gritty; these ashes exhibit dilatancy, are non-plastic and possess cohesion.

Consistency

The particle size distribution of fly ash can vary considerably depending on how a power station is being operated. However, many stations operate some base load generation, reducing the potential variation of the resulting fly ash. Figure 1.14 shows the fineness as the percentage retained on the 45 μm sieve with the date sampled from five sources of fly ash throughout the country. The mean standard deviation in the fineness is 3·6%, with values ranging from 2·9% to 4·4%, remarkably consistent considering the diversity of the sources. Even the higher standard deviation material complies with the fineness criteria contained in EN 450 'Fly ash for concrete'; e.g. 0 to 40 ± 10% retained on the 45 μm sieve.

For use in concrete a considerable proportion of fly ash is classified to make it finer, thereby improving its performance, e.g. BS 3892 Part 1 PFA. The process of classification reduces the variability in fineness significantly, with the standard deviation for fineness ranging between 1·2% and 1·9% with a mean of 1·4%. The fineness with time for classified fly ash,

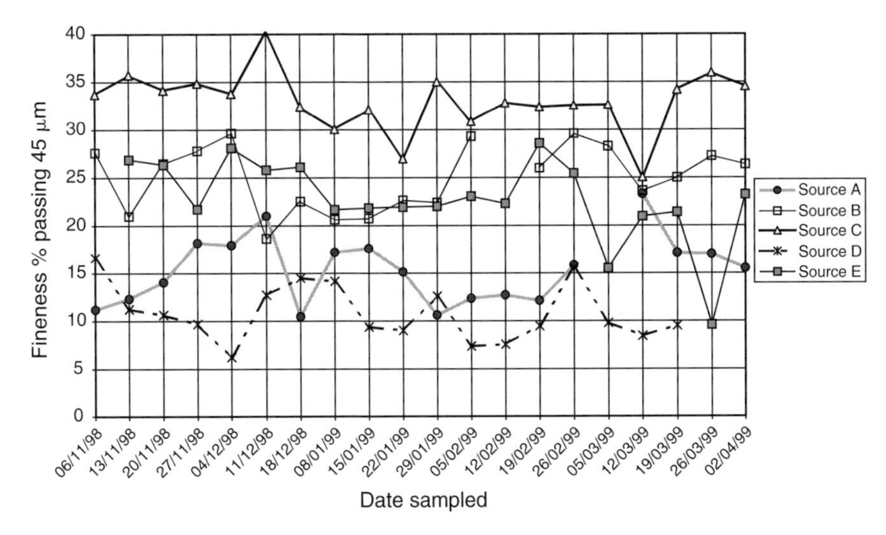

Fig. 1.14. Typical variation in the fineness of fly ash from five UK power stations

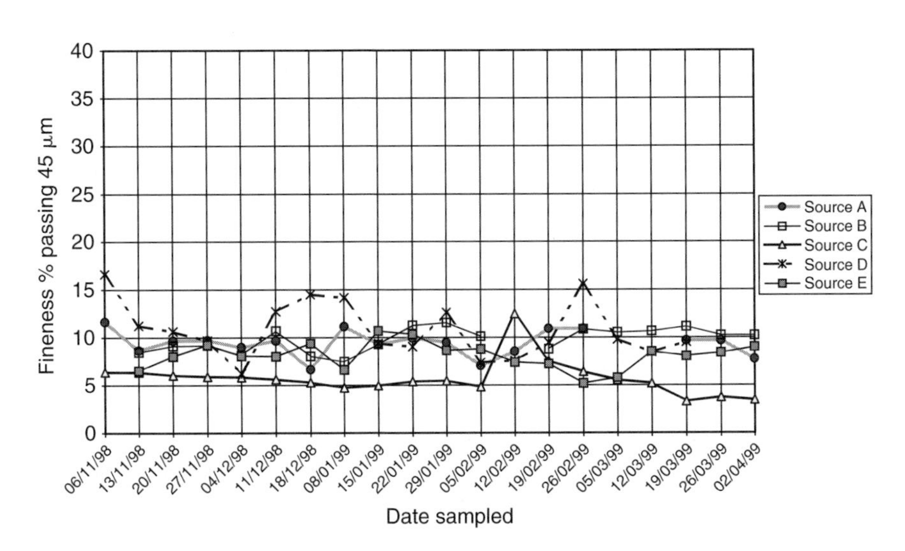

Fig. 1.15. Classification of fly ash reduces the variation in fineness

i.e. PFA for use in concrete, is shown in Fig. 1.15. Fly ash and PFA in concrete are discussed in detail in Chapter 3.

Physical properties

Table 1.4 lists some typical properties of fly ash.

Table 1.4. Some properties of fly ash

Property	Typical value(s)
Compacted bulk density	1200–1700 kg/m^3
Loose bulk density	1100–1600 kg/m^3
Relative density (oven dry)	2·0–2·4
Specific heat capacity	0·8–0·7 J/kg/°C
Water permeability – compacted fly ash	10^{-9}–10^{-6} m/s
Electrical conductivity	0·09 W/mK

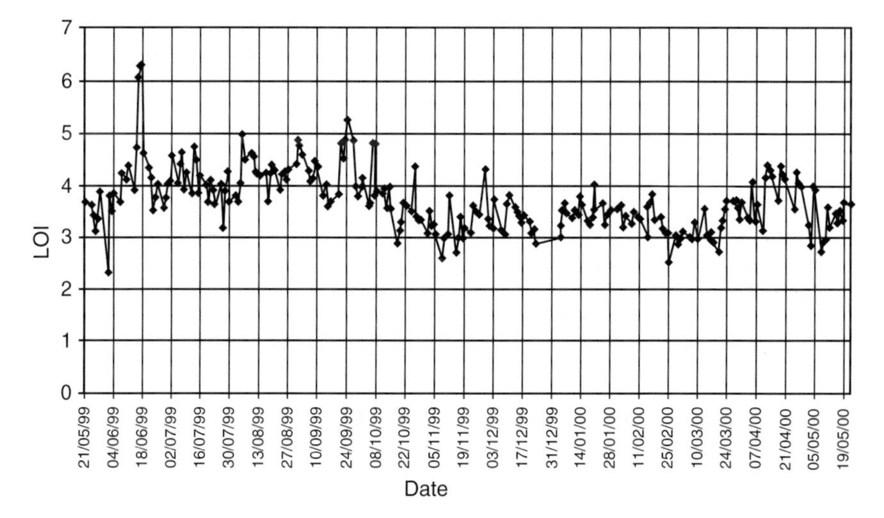

Fig. 1.16. Typical variation in loss on ignition (LOI) with time for fly ash from a large UK power station

Loss on ignition

Carbon content, as assessed by measuring LOI, can vary widely. Before the introduction of low NO_x burners LOI values below 1% were possible, but their introduction led to a gradual rise in LOI to the present levels. When a station is continuously producing electricity, called base loading, the LOI may be typically 3·5%. However, older stations and stations being used to supplement the electricity grid during periods of peak demand, called 'double shifting', may have LOI values >10%. High-rank coals such as anthracite and steam coal can also present combustion problems in the boiler resulting in higher LOI values. The true carbon content of a fly ash is lower than indicated by the LOI. Typically, the true carbon content will be 0·90 of the LOI value. With coal-fired stations double shifting during the summer months there is a tendency for LOI to increase (Fig. 1.16).

Table 1.5. Typical range of analyses from UK fly ash

Element	Typical range of values for fly ash
Silicon (% as SiO_2)	38–52
Aluminium (% as Al_2O_3)	20–40
Iron (% as Fe_2O_3)	6–16
Calcium (% as CaO)	1·8–10
Magnesium (% as MgO)	1·0–3·5
Sodium (% as Na_2O)	0·8–1·8
Potassium (% as K_2O)	2·3–4·5
Titanium (% as TiO_2)	0·9–1·1
Chloride (% as Cl)	0·01–0·02*
Loss on ignition (%)	3–20
Sulfate (% as SO_3)	0·35–2·5
Free calcium oxide (%)	<0·1–1·0
Water soluble sulfate (g/l as SO_4)	1·3–4·0
2:1 water solid extract	
Total alkalis (% as $Na_2O_{eq.}$)	2·0–5·5%
Water-soluble alkalis (% as $Na_2O_{eq.}$)	0·3–1·0%
pH	9–12

*Chloride may be up to 0·3% for fly ash conditioned with seawater.

Chemical and mineralogical properties

Fly ash has three main elements, silicon, aluminium and iron, the oxides of which account for 75–85% of the material. Fly ash consists principally of glassy spheres together with some crystalline matter and unburnt carbon.

Silicon and aluminium are mainly present in the glassy phase, with small amounts of quartz and mullite ($3Al_2O_3$, $2SiO_2$) included. The iron appears partly as the oxides magnetite (Fe_3O_4) and haematite (Fe_2O_3), with the rest in the glassy phase. The greater proportion of fly ash is a glass, and mineralogical examination from six power stations[21] showed that the glass content varied from 66% to 88%. Analysis of the glass content from those six stations showed a remarkable similarity in composition. The $SiO_2 + Al_2O_3$ content varied between 70% and 88% and other constituents included iron, calcium, magnesium, sodium, potassium and titanium.

A typical range of oxides determined by chemical analyses of UK fly ash is shown in Table 1.5.

Trace element analysis

Typical trace elemental analyses are shown in Table 1.6, which demonstrates that other elements are present in only small quantities, <1% of the

Table 1.6. Solid-phase trace element analysis: typical ranges from UK sources of fly ash

Element	Typical range of results	Element	Typical range of results
Antimony	1–325	Fluoride	0–200
Arsenic	4–109	Lead	<1*–976
Barium	0–36,000	Manganese	103–1555
Boron	5–310	Mercury	<0·01*–0·61
Cadmium	<1·0*–4	Molybdenum	3–81
Chloride	0–2990	Nickel	108–583
Chromium	97–192	Phosphorus	372–2818
Cobalt	2–115	Selenium	4–162
Copper	119–474	Tin	933–1847
Cyanide	<0·10*	Vanadium	292–1339
		Zinc	148–918

All data are expressed as mg/kg.
*Below the limit of detection.

total. The values quoted are generally in agreement with other quoted values.[22,23]

The following chapters will explain the many uses to which fly ash can be put.

References

1. *Quality of life counts – Indicators for a strategy for sustainable development for the United Kingdom: a baseline assessment*. Stationery Office, London, Government Statistical Service, 2000.
2. Thernox Unipede. NO_X abatement in coal-fired power stations and the consequences for fly ash quality. *Thermal Generation Study*, Committee 20.03, 01003 Ren 9317, Paris, 1993.
3. Unsworth JF, Barratt DJ, Roberst PT. Coal quality and combustion performance – an international perspective. *Coal and Science Technology*. Elsevier, Amsterdam, 1991.
4. BS EN 196-2. *Methods of testing cement Part 2. Chemical analysis of cement*. BSI, London, 1995.
5. van den Berg JW. Effect of low NO_x technologies on fly ash qualities, *ECOBA/ACCA Joint Meeting*, Toronto, 1998.
6. Robl TL *et al.*, The impact of conversion to low NO_x burners on ash characteristics. *Proceedings of the 1995 International Joint Power Generation Conference*, Minneapolis, October 1995.
7. BS 3892 Part 1. *Specification for pulverised fuel ash for use with Portland cement*. BSI, London, 1997. ISBN 0-580-26785-7.

8. BS EN 450. *Fly ash for concrete – definitions, requirements and quality control.* BSI, London, 1997. ISBN 0-580-24612-4.
9. Brown RD. Ash concrete – its engineering performance. *Ashtech '84,* London, 1984: 295–301.
10. Dhir RK, McCarthy MJ, Tittle PAJ, Kii HH. Use of conditioned fly ash in concrete: strength development and critical durability. University of Dundee, DETR research contract report 39/3/448 (CC 1411), April 2000.
11. McCarthy MJ, Tittle PAJ, Dhir RK. Lagoon fly ash: feasibility for use as a binder in concrete. *Materials and Structures/Materiaux et Constructions* 1998; **31**: 699–706.
12. Emberson PM. The integration of ash disposal in the landscape: some aspects of the UK experience. *Ashtech '84 Conference,* London, 1984: 627–633.
13. Hurt R, Suuberg E, Gao Y. Unburned carbon in ash: formation, properties and behaviour in construction applications, EPRI report TR-109340. *Effects of coal quality on power plants: 5th International Conference,* November 1997.
14. Levy EK. Reduction of fly ash LOI using a bubbling fluidised bed separator. *12th International Symposium on Coal Combustion By-product Management and Use proceedings,* ACCA, January 1997.
15. Stencel JM, Ban H, Li T *et al.* Dry, electrostatic separation of carbon from coal combustion fly ash, *12th International Symposium on Coal Combustion By-product Management and Use proceedings,* ACCA, January 1997.
16. Bittner J, Gasiorowski S, Tondu E, Vasiliauskas A. STI fly ash separation system – operating history of New England Power's Brayton Point power plant, *12th International Symposium on Coal Combustion By-product Management and Use proceedings,* ACCA, January 1997.
17. For more information visit Santee Cooper's website on http://www.santeecooper.com/newsroom/2000releases/ash-6-28.html
18. Matsufuji Y, Kohata H, Tagaya K *et al.* Study on the properties of concrete with ultra fine particles produced from fly ash. *Fly ash, silica fume and natural pozzolans in concrete, Proceedings of an International Conference,* ACI SP132, Vol. 1, Istanbul, 1993: 351–365.
19. Cornelissen HAW. Micronised fly ash – a valuable resource for concrete. *12th International Symposium on Coal Combustion By-product Management and Use Proceedings,* ACCA, January 1997.
20. BS 1377. *Methods of testing for soils for civil engineering purposes, Part 1: General requirements for test methods.* BSI, London, 1990.
21. Central Electricity Generating Board. *Fly ash utilization.* CEGB, London, 1972.
22. Brown J, Ray NJ, Ball M. The disposal of pulverised fuel ash in water supply catchment areas. *Water Research* 1976; **10**: 1115–1121.
23. Hoeksema HW. Working conditions for fly ash workers and radiological consequences of living in a fly ash house. *Proceedings of ASHTech 84, 2nd International Conference on Ash Technology and Marketing,* London, 1984.

Chapter 2

Fly ash and the environment

Introduction

The word 'environment' means '*surroundings; conditions of life or growth*'. This chapter concentrates on the environmental impacts of using fly ash rather than the complex arguments about the combustion of coal, extraction of coal, power-station construction, etc. Fly ash, being a by-product of the generation of electricity from coal, with electricity being the primary product, has no intrinsic impact in respect of global warming gases. However, its unique properties are such that it can be used to reduce the impacts of other industries. Some of the factors worth considering are as follows:

- Many of the applications for fly ash replace naturally occurring aggregates and minerals, e.g. in-fill, road construction and brick-making applications. This can significantly reduce the demand for virgin aggregate.
- By using the pozzolanic reaction fly ash can complement and replace Portland cement in cementitious applications. There is a consequent reduction in greenhouse gas emissions.
- Fly ash can act as a filler extending a material, e.g. the use of cenospheres in plastic (see Chapter 9).

Fly ash, concrete and environmental impact

The manufacture of Portland cement, by the very nature of the constituents and its chemistry, involves the production of a number of so-called greenhouse gases. The main one is carbon dioxide (CO_2), not from the burning of fossil fuels but from the calcining of calcium carbonate to calcium oxide. Approximately 1 tonne of CO_2 is produced for every tonne of Portland cement made. Parrott[1] reports in some detail on the improvements that have been made to reduce emissions in the UK. However, it is clear that there is little room for significant further

Table 2.1. Cement required to manufacture 40,000,000 m^3 of concrete: typical UK annual production (redrawn from Parrott[2])

	PC	GGBS	PFA	Total cementitious
Replacement level	–	50%	30%	–
Extra cementitious	–	5%	10%	–
Tonnes used, with partial replacement of PC by GGBS and PFA	10,560,000	960,000	480,000	12,000,000
Tonnes of PC needed if no replacement by GGBS or PFA	11,763,000	–	–	11,763,000
Savings in PC	1,203,000			

PC: Portland cement; GGBS: ground granulated blastfurnace slag; PFA: pulverised fuel ash.

improvements in the Portland cement-making process without the chemistry being changed.

The replacement of some of the Portland cement offers greater potential as a method of reducing environmental impact. In the UK only two additions fit this category, ground granulated blastfurnace slag (GGBS) and fly ash/pulverised fuel ash (PFA). Parrott[2] produced Table 2.1 based on the 1998 utilisation of GGBS and PFA that shows some 1,203,000 tonnes of Portland cement are being saved annually with current rates of addition utilisation.

On average, a 30% replacement of Portland cement with PFA with suitable adjustment for equal 28 strength reduces overall greenhouse gas emissions by 17%. These reductions in emissions could be fully realised if the use of additions was increased. In principle, a threefold increase in fly ash usage would be possible.

Leachates from fly ash: chemistry and precautions

Composition

A small proportion of fly ash, typically 2–3% by weight, is soluble in water. The soluble form of fly ash is usually alkaline in reaction and mainly contains calcium and sulfate ions. Major elements within fly ash leachate comprise the principal cation of calcium and sulfate in a range $<0{\cdot}1$ g/l. Other water-soluble anion components such as magnesium, sodium and potassium are usually present but in smaller quantities than cations.

Typical trends in the leachate formation process observed in the laboratory have indicated that any initial formation of leachate fly ash is acidic with sulfate ions present as a surface coating. This becomes hydrolysed to form sulfuric acid, resulting in a reduction in pH. Thereafter, the pH rises rapidly, turning alkaline over a comparatively short period of 48 h. Surface deposits buffer pH levels within the leachate, leading to an adjustment in pH, eventually settling out at a residual level within the alkaline range of 9–12. The soluble alkaline ash core is known to deplete; however, this process is slow, may take many years to complete and is not considered to have an impact in terms of leachate formation.

Leachate testing

Historically, the accepted method of determining leachate in the field has been to use laboratory-based tests such as the column and/or lysimeter. These have been used to determine the level of leachate likely to be generated based on certain local geological criteria and rainfall data.

Lysimetric studies have been undertaken by numerous authors who have simulated natural weathering conditions. Lysimeters exhibit a very slow release of leachant.[3] In addition, other methods of generating leachate under laboratory conditions such as those set out by the National Rivers Authority (NRA) and DIN 38414 standards have been evaluated and are considered acceptable methodologies within the UK for assessing leachate quality by the statutory authorities. The data obtained from these methods have been found to compare favourably when modelling the level of leachate dilution likely to take place under site-specific conditions by the Central Electricity Generating Board (CEGB) (Fig. 2.1).[4–6]

Results from column tests[7] have shown that apart from the initial 'flush' of primary surface components, concentrations of cadmium (Cd), chromium (Cr), copper (Cu), iron (Fe), mercury (Hg), manganese (Mn), lead (Pb), titanium (Ti) and zinc (Zn) rapidly fell below detection limits.

Other components such as arsenic (As) demonstrated a delayed release mechanism, whereas a high percentage of boron (B) was shown by Eary *et al.*[8] to be flushed out within the first 48 h. It should be noted that the exact mechanism of release for some elements is not well understood; however, experimental data obtained by the CEGB[9] through column tests agreed well with observed patterns.

At present, evaluation of a Europe-wide test[10] is being undertaken. This has been concentrating on a 10 : 1 liquid to solid phase dilution factor and is similar to the NRA and DIN 38414 tests. This has been the generally adopted factor in considering the likely level of attenuation able to be achieved *in situ* through groundwater flow effects. These tests are for a particle range between <4 and <10 mm. These procedures are intended to form the CEN suite of tests and are envisaged to comprise a mixture of trickle[11] and shake tests.

Fig. 2.1. Barlow mound, Drax, UK, part of which is a site of special scientific interest

It is considered unlikely that these procedures will be found suitable for the testing of monolithic samples. Appropriate applications for fly ashes, especially grouting, often extend particle size distribution beyond the range shown above. The monolithic test[12] is likely to prove a more reliable test and is scheduled for 2004. This test is again anticipated to be to the 10 : 1 dilution rate; however, there is some debate over how a monolithic sample should be defined. It has been proposed that a monolithic test would be suitable for samples taken for a range of compressive strengths and dimension ratios. It is anticipated that a particle size in the order of >40 mm is likely to be appropriate under the proposed monolithic test. Another potential test is being developed, and due in early 2001 before the monolithic test, to consider the use of a larger particle of <32 mm whereby abrasion is avoided and release is therefore controlled by matrix diffusion.

Leaching characteristics

Experiments on the leaching patterns for weathered fly ash within field or laboratory conditions show two prominent patterns. These are comprised of leachate formation, either initially containing high concentrations of calcium (950 mg/l), sodium (400 mg/l), potassium (5 mg/l) and sulfur (5200 mg/l as SO_4), or through a delayed leachate formation leading to high levels of aluminium and silicon being discharged owing to the presence of magnesium and iron.

It has been suggested that high initial concentrations occur as a result of the possible hydrolysis of calcium oxide present in an elevated pH, typically >10, with free calcium species present as sub-micrometre surface particles. High levels of sodium and potassium species found within initial contributions are derived from their oxides with further dissolution of core alkaline oxides and buffering by aluminium oxides contributing to a rise in pH levels to levels indicated earlier. Magnesium solubility is highly pH dependent and has limited solubility at high pH.

Significant trends exist within magnesium species present with the formation of leachate appearing consistent with the chemistry of magnesium oxide after initial low, near-zero concentrations. The release rates of minor elements are determined by the total concentration, their distribution within the ash and their method of final incorporation into secondary (or weathered) solids.

Geochemical processes have been shown to control final concentrations and consist of the following mechanisms:

- precipitation/dissolution
- redox speciation
- adsorption/desorption
- effect of pH upon solubility.

Elements may associate with specific minerals according to the chemistry of the following properties:

- chalcophilic
- lithophilic
- siderophilic.

In response to such concerns regarding the perceived risk of contamination of elements leaching into groundwater, a substantial amount of research work has been carried out. This indicates that leaching is not considered a problem with fly ash compared with levels of trace elements considered acceptable by the statutory authorities.

While it is possible to extract leachate-containing trace metals from fly ash, the conditions under which these can be extracted are not those encountered when the material is used, e.g. in fill applications (Fig. 2.2). The leachability of trace elements depends largely on the quantity and acidity of the percolating water and on the buffering capacity and redox potential of the residues. For most trace elements leachate concentrations increase with increased acidity and with increasing redox potential. Acid-extractable leachate obtained under these conditions contains trace element concentrations far higher than would be obtained with rainwater leaching through alkaline residues under the conditions that exist in fly

Fig. 2.2. Fly ash used as a fill material on the A69 extension, Newcastle upon Tyne, UK

ash fills. In addition, studies have shown that the nitric acid-extractable fraction of trace elements from material landfilled around 20 years ago is virtually identical to that extracted from recently deposited fly ash. This clearly demonstrates that leaching does not occur to any significant extent with rainwater percolation through alkaline residues, since a lower acid-extractable fraction would be expected in the older fly ash if it had been leaching for this period.

Fly ash permeability

The particle size distribution means that compacted fly ash has a low permeability, typically 10^{-7} m/s or lower. This means that it is difficult for water to penetrate. Because water will only flow through saturated material, this will not occur unless the fly ash is placed in areas below water. Experience has shown that if fly ash is subjected to heavy rain it is unusual for saturation to affect the surface beyond the top 50 mm. Even when saturated there will only be a limited rate of flow through the mass of the material.

Laboratory and field measurements of the permeability of fly ash have shown drainage characteristics that range from practically impervious

Fig. 2.3. Fly ash returned to agricultural use

to poor. The lower permeability of fly ash prevents leaching of soluble material from the mass of the compacted fly ash. Consequently, there have been no recorded cases of concrete structures adjacent to fly ash fill suffering from sulfate attack, despite exposure over many years. However, there may be a need to consider hydraulic effects such as pore pressure distributions in the design of backfilled structures.

Historical work conducted by the former CEGB over many years has consistently demonstrated that leachate from fly ash does not represent a threat to groundwater quality. Ray and Ball[13] studied lagoons and landfills used in the disposal of fly ash in water-supply catchment areas and concluded that if correctly managed these did not present a risk to water supplies for drinking or any other purpose (Fig. 2.3).

Waste labelling and risk assessment

In the latest version of both the EC European Waste Catalogue and the UN–ECE list of wastes both furnace bottom ash (FBA) and fly ash are classified as non-hazardous. In addition, fly ash is classified as non-hazardous in both The Netherlands and the USA, countries that are acutely concerned about environmental impact arising from landfill.

As already stated, around 66–88% of fly ash is present as an amorphous glassy material composed of silica, alumina and iron oxides, with other metals present in smaller quantities. The constituents, apart from the glass, that are of most significance to the properties of fly ash are the calcium oxide content (lime) and sulfate contents.

If there is sufficient lime present in the fly ash then it will result in further hardening due to a combination of further crystal formation and reaction between the lime and the glassy material in the fly ash (pozzolanic reaction). The high pH is likely to reduce the availability of the trace elements.

When water is added to fly ash, it initially has a low pH as the sulfate deposited on the surface of the particles is brought into solution as sulfuric acid. This is a transient situation and the pH rapidly rises as calcium is leached into solution. The pH is typically 9–11 for fly ash, although the pH for those ashes with higher free calcium oxide contents can rise to 12. Only a very small quantity of free calcium is required to achieve the higher pH. Because most of the water-soluble material that influences pH has been washed out of lagoon fly ash, the pH is lower, typically around 9.

The calcium content of fly ash means that most of the sulfate is present as gypsum, which has a limited solubility and will precipitate out in compacted fly ash. The sulfate level of lagoon fly ash is usually very low because the water/solids ratio used to slurry the fly ash means the majority of the sulfate is washed out. Other water-soluble materials are also removed in the process. The sulfate content is typically $<0{\cdot}1$ g/l.

The sulfate content of fly ash means that it cannot be placed within 500 mm of metallic items, according to the UK Department of Transport Specification for Highway Works (SHW). The water-soluble content of fly ash is also sufficiently high to restrict the types of reinforcement that can be used in reinforced earth structures.

Leachable elements

DIN leaching test

As discussed above, only a small fraction of the constituents present on the surface of fly ash is leachable in water. Typical data obtained from routine analysis are shown in Table 2.2; the extraction in this instance is to the German standard DIN 38414-S4[14] (10 : 1 water/solids ratio). This DIN extraction test is very similar to the proposed European test 'Compliance for leaching granular materials and sludges – Part 2: One batch test as a liquid to solid ratio of 10 l/kg with a particle size below 4 mm (with or without size reduction)'; prEN 12457 Part 2.

From Table 2.2 it can be seen that the major water-soluble constituents are calcium and sulfur (usually present as sulfate). There are smaller amounts of sodium and potassium, and traces of chloride, magnesium, aluminium and silicon. If it is assumed that all of the water-soluble calcium, sodium and potassium is present as hydroxide (ignoring the sulfate or chloride) then the total water-soluble hydroxide, based on the

Table 2.2. Leachates found using the DIN 38414-S4 method: typical ranges from UK sources of fly ash

Element	Typical range of leachable elements	Element	Typical range of leachable elements
Aluminium	<0·1*–9·8	Magnesium	<0·1*–3·9
Arsenic	<0·1*	Manganese	<0·1*
Boron	<0·1*–6	Molybdenum	<0·1*–0·6
Barium	0·2–0·4	Sodium	12–33
Calcium	15–216	Nickel	<0·1*
Cadmium	<0·1*	Phosphorus	<0·1*–0·4
Chloride	1·6–17·5	Lead	<0·2*
Cobalt	<0·1*	Sulfur	24–510
Chromium	<0·1*	Antimony	<0·01*
Chromium VI	<0·1*–1	Selenium	<0·01*–0·15
Copper	<0·1*	Silicon	0·5–1·5
Cyanide	<0·01*	Tin	<0·1*
Fluoride	0·2–2·3	Titanium	<0·1*
Iron	<0·1*	Vanadium	<0·1*–0·5
Mercury	<0·01*	Zinc	<0·1*
Potassium	1–19	pH	7–11·7

Data are expressed as mg/l.
The data include a seawater-conditioned sample, hence the high chloride values.
*Value below detection limit.

highest values from Table 2.2, would be 2·1% (m/m). However, calcium hydroxide would make up approximately 2·0% and the other compounds represent <0·1%. In all instances quoted the calcium is highly dominant, with sodium and potassium present in comparatively very small quantities.

Harwell test

Samples[15] of stockpile fly ash were subjected to extraction by the Harwell[16] method and the results are summarised in Table 2.3. Although it is stockpile fly ash, the leachate still shows that the calcium content is dominant, with smaller amounts of other elements. There is a more significant amount of magnesium, probably due to the low pH of these samples.

The CIRIA report[17] compared the CEN two-stage leaching method to the Harwell method. It demonstrated that the total leachate was similar for both techniques for the three fly ashes examined. Both showed that with the Harwell method the calcium and sulfate were present in the range of 100–1000 or 1000–10,000 mg/l, respectively, for the first bed volume extracted, but both fell dramatically to the 100–1000 mg/l range for

Table 2.3. Leachates found using the Harwell method: typical range from 10 samples from a single UK source of fly ash

Element	Typical range of leachable elements	Element	Typical range of leachable elements
Bed volume	1	Molybdenum	0·15–0·88
pH	8·1–8·8	Sodium	5–44
Aluminium	<0·1*–0·5	Nickel	<0·01*
Arsenic	0·06–0·16	Lead	<0·01*
Boron	1·8–4·3	Tin	<0·01*
Calcium	33–250	Titanium	<0·01*
Cadmium	<0·005*	Vanadium	0·22–0·55
Cobalt	<0·01*	Zinc	<0·01*
Chromium	0·02–0·06	Nitrogen	0·2–1
Copper	<0·01*	Phosphorus	<0·1*
Iron	<0·01*	Sulfur	15–70
Mercury	<0·001*	Chlorine	5–9
Potassium	5–29	Fluorine	<0·1*
Magnesium	16–100	Selenium	0·04–0·16
Manganese	<0·01*	Antimony	0·01–0·02

Data are expressed as mg/l.
*Value below detection limit.

the next four bed volumes extracted. The concentration for the sixth to tenth bed volumes fell to 10–100 mg/l. The only other elements that were present at these levels were sodium and potassium, but these had significantly reduced by the third bed volume for all but one of the fly ashes examined.

Leachates from fly ash: summary

1. The majority of the ash is present as an alumino-silicate glass.
2. Most elements are present in very small quantities and are largely entrained in the glassy material.
3. Typically <2% of the fly ash is water soluble; calcium and sulfate constitute the majority of the water-soluble fraction. There are smaller amounts of sodium, potassium and, in low pH leachate, magnesium.
4. The pH is mainly determined by the water-soluble calcium and sulfate.
5. The water-soluble fraction, although small, can be sufficient to produce a pH above 11·5, but dilution can rapidly reduce the water-soluble fraction and therefore the pH.

Polycyclic aromatic hydrocarbons in fly ash

Polycyclic aromatic hydrocarbons (PAHs) can result from the incomplete combustion of fuels such as wood, coal and oil. Metabolic transformations, by aquatic and terrestrial organisms, result in carcinogenic substances.[18] The most potent PAHs are benzofluoranthenes, benzo[a]pyrene, benz[a]-anthracene, dibenzo[a,h]anthracene and indenol[1,2,3-cd]pyrene. Although there has been a significant amount of work on PAHs arising from combustion of coal, most effort has focused on airborne particulate matter. PAHs will undergo photodegradation and are therefore thought to have a limited lifespan in the atmosphere. PAHs are only sparingly soluble in water, but their solubility decreases with increasing size of the molecule; e.g. naphthalene with two benzene rings has a solubility of 32 mg/l whereas benzo[a]pyrene with five rings has a solubility of $1{\cdot}6 \times 10^{-3}$ mg/l.

Wild and Jones[18] reviewed PAHs in the environment in 1995. The major sources, apart from gasworks sites, were found to be coal-fired electricity generation (3140 tonnes per annum), domestic coal combustion (600 tonnes per annum), incinerators (56 tonnes per annum) and vehicles (80 tonnes per annum), with smaller amounts from oil and wood combustion and stubble burning. Sharkey *et al.*[19] noted that PAHs could also arise from a number of other sources, such as coke ovens and metal smelting and processing plants. Although power generation consumes the largest proportion of coal, the emissions of PAHs are less per tonne than for coal used for domestic consumption. This is due to the higher temperatures and greater control over combustion conditions in the former. It was noted also that concentrations of PAH were higher in urban areas and by the roadsides, as might be expected. When deposited in water, PAHs tended to bind to sediments rather than remain in solution: the estimated burden of PAHs in sediments was 10 times greater than in freshwater. Levels in the soil are rising as a result of atmospheric deposition and binding of the PAHs by the soil particles, especially the organic matter. The report considered that leaching of PAHs from soils was unlikely to be significant.

Wright,[20] working for the CEGB, indicated that the amount of PAHs present in fly ash is small, typically in the range 141–935 ng/g. However, it was noted that the amount detected was likely to be a small fraction of the total owing to difficulties in extracting them. The conclusion was that the quantities found and the inert nature of fly ash meant that fly ash was not a major contributor to 'active' PAH in the environment.

Another study[21] by Zenon Environmental detected no extractable trace organics in fly ash, from large coal-fired power stations, in concentrations above 0·5 μg/g, similar levels to those found by the CEGB.

Leaching tests on fly ash in accordance with the Environment Agency extraction method[22] have indicated levels of the PAHs benzo[b]fluoranthene, benzo[k]flouranthene, benzo[a]pyrene, benzo[ghi]perylene, fluoranthene and indeno[1,2,3-cd]pyrene to be <0·2 μg/l for each species, confirming the above findings that the amount of available PAH from the fly ash is negligible.

There have been a number of studies in the USA investigating the interactions between PAHs and fly ash.

One recurring theme is the difficulty in extracting PAHs from particulate matter. Janssen and Kanij[23] carried out desorption tests using ^{14}C-labelled 3,4-benzopyrene (BaP). They found that recovery from fly ash was lower than for aluminium powder of a slightly greater surface area. Extraction using xylene was found to give the lowest recovery of BaP. Tests on a number of different fly ashes gave recoveries up to 50%, compared with up to 95% for aluminium powder. Thermal treatment of the fly ash at 400°C reduced recovery significantly, the lowest recovery being 8%.

Natusch[24] found that PAHs tended to adsorb on to the surface of fly ash, rather than just being a surface deposit. It was noted that adsorption was very temperature dependent, with little adsorption on stack ash at 290°C but significantly more at 5°C. The adsorption occurred rapidly but it was difficult to desorb PAHs. Adsorbed PAH is resistant to photodegradation although some oxidation may occur at a lower rate on ash that has not been exposed to light. The behaviour of PAHs with respect to fly ash was found to be similar to that with activated carbon. The report noted that the bulk of the PAHs not adsorbed by carbonaceous matter would be adsorbed on the fine particles owing to the large surface area, these particles being most likely to become airborne.

Harrison *et al.*[25] studied both solid and liquid wastes from Four Corners power station in New Mexico, USA. They found that recovery of PAHs from fly ash was difficult and they did not detect any PAH with more than four rings, although they could not confirm that these were not present on the surface of the ash. The levels in precipitator ash were slightly higher than for ash collected by the water scrubbing system. Total extractable hydrocarbons in sluice water were low, at 7·9 ppb, with many species at levels <1 ppt. No PAHs larger in molecular weight than naphthalene were found in sluice waters.

Junk *et al.*[26] studied levels of PAHs and other organics from stack vapour, stack ash, fly ash and grate ash from Ames power station in the USA. Only small amounts were found on both respirable and non-respirable particles (Table 2.4). Although there were measurable amounts in the vapour phase, it was noted that if all the vapour were to condense on the particulate matter the amount would still be less than for ambient air particles.

In addition to the above, measurements were made on sluice water carrying fly ash and grate ash to settling ponds. The water used was originally from an aquifer that had been contaminated by coal tar and

Table 2.4. Levels of polycyclic aromatic hydrocarbons in emissions from coal-fired power stations (Junk et al.[26]*)*

Compound	Concentration range (ng/g)		Concentration range (ng/m^3)
	Respirable particles	Non-respirable particles	Vapour phase
Naphthalene	ND–18	0·5–23	10–1800
Phenanthrene	–	–	26–640
Anthracene	–	–	0·4–100
Fluoranthene	0·2–0·3	0·05–1·5	0·5–240
Pyrene	0·2–7	0·08–1·1	0·2–2850
Chrysene	ND	ND–4	0·1–28
Benz[a]pyrene	ND	ND	0·1–120
Benz[a]anthracene	ND	ND–0·3	NM
Benz[ghi]perylene	NM	NM	3–22

ND: not detected at the limit of 0·05 ng/g; NM: not measured.

contained some PAHs. The sluice water, having been in contact with the fly ash and grate ash, contained no contaminants above the detection limit of 1 ppb, i.e. less than found in the aquifer water. This indicated that the fly ash reduces the level of PAHs in the water. This was confirmed by a small trial where water containing 20–50 ppb of PAHs was mixed with fly ash in a ratio of 10:1. Within 10 min the PAH level was reduced to below the detection limit.

In 1989 Mamantov and Wehry[27] separated fly ash into carbonaceous, magnetic, light mineral and heavy mineral fractions. They reported that PAHs with three or four rings may be found in equilibrium in the atmosphere both as adsorbates on particulate surfaces and in the vapour phase. Larger PAHs were present as adsorbates. Although PAHs will photodegrade, the extent to which this occurs is dependent on the surface on which they are adsorbed, so they may be more persistent on some surfaces than on others. It was noted that resistance to degradation was increased when the PAHs were adsorbed on carbonaceous and magnetic fractions. Their study used a vapour deposition technique instead of the more usual solution-based techniques to deposit pyrene on to the surface of fractions from two sources of fly ash. This was considered to be a more realistic method of deposition of the PAHs. They reached the following conclusions.

- Pyrene had a greater affinity for the carbonaceous fraction of the fly ash than the other fractions.
- In the absence of carbonaceous material pyrene preferentially adsorbs on the non-magnetic fraction.

- Pyrene adsorbed on the carbonaceous fraction was very resistant to photodegradation.
- Although the pyrene is not readily adsorbed on the magnetic fraction, any that is will be resistant to photodegradation.
- Carbonaceous matter tends to be in the coarser fraction of the fly ash but it is the finer respirable material that may result in atmospheric emissions via the stack.

Further work by the same authors was reported in 1995.[28] This involved trying to ascertain the effect of surface roughness on photodegradation. By measuring the surface area of the various fractions, they showed that the carbonaceous fraction had the greatest influence on the surface area. In one sample the carbonaceous fraction represented 4·7% of the material by mass but accounted for 80% of the measured surface area.

The work confirmed that the surface roughness was important and that the iron content, which tends to produce dark-coloured particles that may absorb light, had little effect.

Polycyclic aromatic hydrocarbons in fly ash: summary

1. Most work has been done on PAHs associated with particulate matter that is released to the atmosphere. Little work has been done on water-borne PAHs.
2. There are traces of PAH present on fly ash, typically up to 900 ng/g, although the difficulty in recovering PAHs from fly ash means that the figure may be higher.
3. PAHs, particularly those with high molecular mass, tend to be adsorbed on to the surface of the fly ash and recovery of the adsorbed PAHs is very difficult.
4. PAHs have an affinity for particulate matter, especially carbonaceous matter, and dissolved PAHs can be removed from solution by fly ash.
5. The leachate from fly ash contains very small amounts of PAHs.
6. PAHs adsorbed on to fly ash tend to be resistant to photodegradation, although it is possible that PAHs not exposed to light can still undergo oxidation.

The concentration of effort on airborne emissions rather than water-borne PAHs indicates the relative significance of each. When fly ash is used as a fill, the amount that may become airborne is small, especially if dust suppression is effective, and transient, occurring only during construction. This means that this is not likely to be a major source of PAHs in the air.

Fly ash, when used as a fill material (Fig. 2.4), will be well compacted and therefore have a low permeability, typically 10^{-7} m/s. Furthermore, if

Fig. 2.4. Fly ash used as a fill material, A52, Derby, UK

used in road embankments it will be protected by topsoil/vegetation as well as the construction of the road. Thus, water movement through the fly ash will be slow to non-existent. Given the difficulty in extracting PAHs from fly ash and their low permeability, the risk of their moving from the fly ash fill is very small.

Furthermore, as discussed by Wild and Jones,[18] when fly ash is used in a road embankment the emission of PAHs from vehicles is likely to be a greater threat than any leaching of PAHs from the fly ash.

Dioxins in fly ash

Polychlorinated dibenzo-*p*-dioxins (PCDDs) are a family of chemicals based on the tricyclic molecule benzo-*p*-dioxin, which has two benzene rings linked by two oxygen atoms. PCDDs have some or all of the hydrogen on the benzene rings replaced by chlorine (up to the maximum of eight chlorines). These are often associated with polychlorinated dibenzofurans (PCDF), which have only a single oxygen atom. Dioxins and furans are considered to be toxic to humans, although furans less so than dioxins. 2,3,7,8-Tetrachlorodibenzo-*p*-dioxin (TCDD) is considered to be the most toxic dioxin and therefore the most studied.

Dioxins are usually associated with the incomplete combustion of material containing chlorine and as such are commonly associated with the ash from municipal waste incineration, but can be found in small traces in soils. The low chlorine content of coal combined with the high temperatures found in the furnaces of power stations mean that dioxins are unlikely to form and only traces would be expected in the resulting ash. Dioxins are ubiquitous and are present in a wide range of soils. Although they can be persistent, they rapidly decay when exposed to light.

Work by the CEGB[29] in the 1980s examined 18 fly ash samples from a range of sources for dioxins from the tetrachlorinated to the octachlorinated. The findings were that the levels were very low, typically <25 pg/g, with levels of 2,3,7,8-TCDD less than 2 pg/g in all but two samples. The only exceptions were samples of fly ash from the low NO_x burners at one station (station A). It was thought that the low NO_x burners might have had some effect, although the same increase was not observed for samples from other power stations fitted with similar burners. Although the dioxin levels in the samples from low NO_x burners at station A were higher, 210 and 270 pg/g, they were still within the range found in soils in the UK. Data from unpublished work cited an upper limit in soils of 290 pg/g.

A sample of cenospheres ('floaters') from one station was sent for analysis in 1993. The analysis included the 17 most significant dioxins and furans with the result quoted as a toxic equivalent (TEQ), relating the total concentration of the 17 species to the concentration of 2,3,7,8-TCDD with equivalent toxicity. This involves applying a weighting factor, the toxic equivalent factor (TEF), to each dioxin or furan, the factor being consistent with its perceived toxicity; the individual results for each species are added together to obtain the TEQ. The highest factor (1) is for 2,3,7,8-TCDD, the lowest is 0·001 for OCDD and OCDF. The results are shown in Table 2.5.

The TEQ is shown as 65 pg/g, which is slightly higher than found in soils (10–40 pg/g) using this method of assessment. However, the density of the floater particles is low compared with soil, with a density of 0·5 mg/m^3 compared to 2·6 mg/m^3. If the value is corrected to an equivalent density then the value would be 12·5 pg/g, similar to the background level in soils and in agreement with the earlier data.

Junk *et al.*[26] looked at 2,3,7,8-TCDD at a detection limit of 10 ppt. No TCDD was found in the effluents of any of the boilers at the two power stations tested. Even when refuse-derived fuel was added no dioxins were observed. This was explained by the high furnace temperature (~1100°C) and the excess oxygen used in combustion.

Dioxins in fly ash: summary

1. Various researchers have confirmed that no dioxins over 25 pg/g are generally found in the ashes from power stations.
2. Although dioxins are present in fly ash, the levels are very low and similar to the background levels found in typical soils. Thus, fly ash is no more hazardous than soil.

Table 2.5. Results of tests on cenospheres

Dioxin/furan	Concentration	TEF	Typical background level in soils (pg/g)
2,3,7,8-TCDF	3·0	0·1	
Total TCDF	30		<0·5–237
2,3,7,8-TCDD	2·0	1	<0·5–2·1
Total TCDD	15		<0·05–69
1,2,3,7,8-PCDF	5·0	0·05	
2,3,4,7,8-PCDF	5·0	0·5	
Total PCDF	35		<0·5–185
1,2,3,7,8-PCDD	2·0	0·5	<0·5–2·4
Total PCDD	10		<0·5–165
1,2,3,4,7,8-HxCDF	10	0·1	
1,2,3,6,7,8-HxCDF	2·0	0·1	
1,2,3,7,8,9-HxCDF	1·0	0·1	
2,3,4,6,7,8-HxCDF	7·0	0·1	
Total HxCDF	35		4·3–212
1,2,3,4,7,8-HxCDD	40	0·1	
1,2,3,6,7,8-HxCDD	50	0·1	
1,2,3,7,8,9-HxCDD	60	0·1	
Total HxCDD	350		2·8–165
1,2,3,4,6,7,8-HpCDF	110	0·01	
1,2,3,4,7,8,9-HpCDF	20	0·01	
Total HpCDF	175		1·5–138
1,2,3,4,6,7,8-HpCDF	1000	0·01	
Total HpCDF	2000		7·5–234
OCDF	6000	0·001	<2·0–144
OCDD	25,000	0·001	28–832
TEQ	65		10–40

TEF: toxic equivalence factor; TEQ: toxic equivalent.

Thresholds for leachates

Table 2.6 shows the Drinking Water Inspectorate thresholds for leachates, PAHs and similar. These are provided for comparison with the various trace compounds found in the sections above.

The leachate quality threshold is applied to leachates from materials in contact with drinking water. The total concentrations data refer to the maximum permitted within a sample of the material. However, normally a dilution factor of 10 : 1 is allowed for leachates as being more representative of the true environmental risk. Barring the initial flush of elements, these leachates would quickly reduce, for example for concrete in contact with drinking water.

Table 2.6. Drinking water threshold values used in the UK (derived from Guidance Note ICRCL Note 59/83)

	Leachate quality threshold (μg/l)	Total concentrations (mg/kg air-dried sample)	
		Lower threshold concentration	Upper threshold concentration
pH	5·5–9·5	6–8	5–9
Toluene extract	–	5000 (subject to special waste)	10,000 (subject to special waste)
Cyclohexane extract	–	2000 (subject to special waste)	5000 (subject to special waste)
Conductivity	1000 μS/cm	–	–
Chemical oxygen demand (COD)	30 mg/l	–	–
Ammonia	0·5 mg/l	–	–
Arsenic	10 μg/l	10 mg/kg	40 mg/kg
Cadmium	1 μg/l	3 mg/kg	15 mg/kg
Chromium (total)	50 μg/l	600 mg/kg	1000 mg/kg
Lead (total)	50 μg/l	500 mg/kg	2000 mg/kg
Mercury	1 μg/l	1 mg/kg	20 mg/kg
Selenium	10 μg/l	3 mg/kg	6 mg/kg
Boron	2000 μg/l	3 mg/kg	–
Copper	20 μg/l	130 mg/kg	–
Nickel	50 μg/l	70 mg/kg	–
Zinc	500 μg/l	300 mg/kg	–
Cyanide (complex)	–	250 mg/kg	250 mg/kg
Cyanide (free)	50 μg/l	25 mg/kg	25 mg/kg
Sulfate (SO_4)	150 mg/l	2000 mg/kg	2000 mg/kg
Sulfide	150 mg/l	250 mg/kg	250 mg/kg
Sulfur (free)	150 mg/l	5000 mg/kg	5000 mg/kg
Phenol	0·5 μg/l	5 mg/kg	5 mg/kg
Iron	100 μg/l	–	–
Chloride	200 mg/l	–	–
Polycyclic aromatic hydrocarbons (PAH)	0·2 μg/l	50 mg/kg	1000 mg/kg

Within the UK, permission to use a material is given by the Environmental Agency and/or Drinking Water Inspectorate on a contract-by-contract basis. Often an environmental risk analysis will be carried out. This implies that for each job a complete reanalysis of the leachates would be required. In practice, acceptance of concrete comprising of cement, fly ash, GGBS, natural aggregates and some admixtures has been accepted as being usable when in contact with drinking water.

Environmental impacts of using fly ash in applications

Fly ash has been successfully used in many applications for many years and there has always been a need to consider its environmental impact. The increasing awareness of environmental issues in recent years has had an impact on sales. With the UK requirement to assess its use on a site-by-site basis, followed by a risk assessment to be carried out, the time taken to obtain approval for the use of fly ash can exceed the time scales imposed by site operations. There has therefore been a need for an alternative approach to the problem. At the time of writing an 'Environmental Code of Practice'[30] has been proposed and put forward to the UK Environment Agency as one solution to this problem.

The primary need to protect the environment is one using a common-sense approach. The following practical considerations must be made when using fly ash.

Recovery of fly ash from stockpiles and lagoons

The recovery process involves exposing an area of stockpile, which is then excavated and loaded into wagons. There is clear potential for dust blow from the exposed face as well as in the handling process. In addition, the removal of some material on the wheels and bodies of the lorries, which may be deposited on the roads, can be a problem. Items such as spillage of diesel and tramp materials need to be considered at the design stage for the fly ash stockpile, but this is not considered within the code of practice for fill.

Similar considerations apply to lagoons as to stockpiles (Fig. 2.5). There may be a greater tendency for the wetter material to adhere to vehicles.

Fig. 2.5. A restored fly ash lagoon

Conditioned fly ash

It is essential that mixers that are used to condition the material, i.e. moisten the fly ash, are designed in such a way to prevent fugitive dust from being emitted from them. Consideration needs to be given to methods of cleaning mixers and disposal of the waste, as well the material deposited on the ground around the loading point.

Dry fly ash

Dry fly ash should be treated in the same way as cement. Sealed silos fitted with filtration units to prevent dust release are required. To maintain fluidity of the material dry aeration is needed. For silos where the material may be stored for some period of time, some form of agitation may be needed to prevent the fly ash from forming a dense mass at the peripheries of the silo. Such devices may consist of air cannons or mechanical stirring systems.

For all the above categories of fly ash, should a vehicle be overloaded, a properly managed procedure for tipping off any excess material is required.

Transport

For moistened fly ash, problems are often associated with dust blow from unprotected wagons and deposition of materials on public roads from the body of the wagon. In order to prevent dust blow during transport all vehicle loads should be sheeted. For dry fly ash, sealed containers such as cement tankers are required. However, surface treatments that stabilise dusty material exist that are suitable for fly ash in some circumstances.

Design of fill structures

Account must be taken of the environmental impact of the construction in the design process for structures using fly ash as a fill material. This will involve ensuring that there is an adequate drainage layer to prevent capillary rise and saturation of the fly ash. In addition, the profile of the fly ash should be such to allow efficient run off rainwater both during and after the construction period. Long-term protection of any side slopes, such as top soiling, is required to prevent build-up of run-off and subsequent potential environmental problems. Suitable methods for encouraging the growth of plants and trees or physical barriers must be designed into the structure.

Laying and compaction of fill

The primary consideration when placing fly ash is one of minimising dust blow by ensuring that fly ash as delivered and after compaction is kept sufficiently moist to prevent dust being created. Windy conditions result in the greatest risk of the fly ash being dried out. Therefore, water-spraying equipment is needed to remoisten the surface. Fly ash that has been accidentally overmoistened can be allowed to dry out by breaking up the surface to encourage evaporation. After a suitable period, the fly ash may be reused when the optimum moisture content is achieved.

If material is stockpiled on-site, it should be deposited in such a way as to prevent accidental contamination of adjacent watercourses. Vehicles leaving the site need to be in a clean condition and provision of wheel washers or similar may be required.

Leaching from cementitious systems

Other than fill applications, fly ash is used in a variety of cementitious systems, e.g. concrete, grouts and sprayed concrete. Cementitious systems are generally excellent at reducing leachates from their constituent components. As fly ash is a pozzolanic material and reacts with the alkalis from Portland cement, the resulting hydration structure becomes very impermeable (see Chapter 3). The low permeability prevents trace elements from the fly ash, the Portland cement and any other constituents from being easily leached. The CEN report[31] on leaching methods from concrete incorporated a number of fly ash concretes. This concluded that 'concrete with bituminous coal fly ash does not show significant leaching of trace elements shown by the fact that for most elements the contents in the leachate were below the detection limit'. Work carried out for the UK Drinking Water Inspectorate[32] also concluded that concrete, including that using fly ash, presented no significant risk for use in contact with drinking water. Exceptions were for concretes that included some types of corrosion-inhibiting admixture.

Agricultural applications

When fly ash is used for fill applications or on disposal sites the normal recommendation is to cover the surface with soil. Woolley *et al.*[33] reviewed the environmental aspects of fly ash in some detail. Fly ash contains no organic matter or clay minerals to hold nutrients and when fresh is sterile. It is deficient in nitrogen, and although phosphorus levels appear significant, the high levels of aluminium oxides and the high pH prevent phosphorus uptake. However, potash and sulfur are plentiful.

Soluble boron within fly ash that has not been lagooned causes some toxicity problems. Boron content, when extracted using boiling water, ranges from 5 to 200 mg/kg depending on coal type and source. In comparison, the normal range for UK soils is 0·4–3·1 mg/kg. Approximately half of the total boron is available. However, because of the natural presence of boron in some soils and irrigation waters, crops have been categorised into

Fig. 2.6. Tree growth on a fly ash fill embankment after 35 years

boron-tolerant and -intolerant species. Tolerant species include perennial rye grass, clovers, mustard, carrot, sugar beet, beetroot and spinach. Sweet clover is the most suitable first crop on fly ash owing to its relatively high uptake of boron. Subsequent weathering also reduces the boron available to plants.

Molybdenum has been identified as a potential problem for ruminant animals, e.g. cattle and sheep. A high molybdenum intake interferes with the animals' copper uptake, which could lead to growth problems and a syndrome called 'teart'. Natural soils can have similarly high molybdenum levels. However, by increasing the copper intake for such ruminants these problems can easily be overcome.

Exposed, untreated fly ash stockpiles will establish mosses, followed by grasses, eventually willow and birch trees over a 30-year period (Fig. 2.6). Orchids particularly like fly ash as a growing medium because of the low fertility due to a lack of nitrogen and phosphorus.

Summary of environmental impacts

It has been established that the leachate is small and consists mainly of calcium sulfate. Only very small amounts of other elements are available to leach, the most significant being boron because of its potential problems for plant growth. However, these problems can be overcome by the appropriate selection of plants. Furthermore, there is little significant difference in leachate quality between all UK fly ashes, and these are well established. Lagoon fly ash will have lower leaching potential, the leachates being removed during the lagooning process. Therefore, only limited testing is required to demonstrate consistency of the leachate, although if there is any major change in fuel burnt then there will be a need to demonstrate that there is no effect on leachate quality.

The typical construction of a fly ash embankment is to place a drainage layer beneath the fly ash to prevent capillary rise, to protect the side slopes with topsoil and/or vegetation and to build the road on top. In such situations the fly ash is effectively isolated from the surrounding environment and therefore does not present a risk with regard to pollution. The major risk of water ingress would be from cracked surfacing, but this would be for a limited period and the effect would be small owing to the impermeability of the fly ash. Thus, if such structures are to be built there is no need for a major impact assessment other than establishing that the fuel burn, and therefore leachate potential of the fly ash, is consistent, and providing a method statement for handling the material on site. The method statement should cover measures to prevent accidental discharge into watercourses and prevent dust blow.

In other types of fill structure, where the fly ash cannot be protected to the same extent, an environmental impact assessment may be required. This should take into account the low permeability of the fly ash. There is again no need for significant amounts of testing provided the producer can demonstrate the consistency of the leachate potential.

If monitoring of the fly ash is required for any reason then the nature of the monitoring needs to be established. Both conductivity and pH measurement can be readily measured.

- There are few, if any, problems with leachates, e.g. small amounts and low permeability.
- All UK fly ash sources appear similar.
- There are no known problems of leachates causing environmental problems, in over 40 years of usage.

Therefore, the following procedures are recommended:

- Leachates should be monitored annually at the point of supply using an approved method from each source to ensure consistency of material.
- No further testing should be required unless there is a major change in fuel type.
- If fly ash is to be used within an environmentally sensitive area then conductivity and pH monitoring can be carried out at the rate agreed between the customer and the supplier.

Radioactivity

Radiation from fly ash results from the concentration of natural minerals within the coal, e.g. the carbon fraction is removed when fired. The nat-ural radioactivity of both the coal and the ash results mainly from the radionuclides from the decay series of uranium and thorium, as well as potassium-40. There is no significant increase in the radioactive composition as nothing is added or no process used that could cause such an increase. Potassium-40 decays into calcium-40 or argon-40, both of which are stable nuclides that will not decay further. From coal ash the K_{40} content of potassium is only 0·012% and from the radiation viewpoint K_{40} has little significance.

In 1986 the CEGB commissioned a comprehensive study of the emissions from UK fly ashes, as reported by Green.[34] The project included fly ash, building materials made using fly ash and field studies of radiation from buildings and ash disposal sites. Table 2.7 summarises the data from this work.

Table 2.7. Summary of estimates of annual effective dose equivalents from fly ash (Green[34])

Situation	Annual effective dose equivalent					
	Normal ground			Fly ash disposal site with 500 mm of soil cover		
	From γ	From R_n	Total	From γ	From R_n	Total
Indoors						
All-brick dwelling	740	260	1000	750	360	1110
Heavy block dwelling	700	290	990	710	400	1110
Light block dwelling	530	340	870	540	440	980
Outdoors						
Workers such as farm	56	57	110	70	60	130
or disposal site						
labourer (2000 h p.a.)	14	7	21	18	8	26
Members of the						
public (500 h p.a.)						
Inhalation of			11			
resuspended dust						
(8760 h p.a.)						

Values are rounded to two significant figures.

Green concluded that

- the incorporation of fly ash in building materials results in increased radiation exposure compared with the use of traditional clay bricks
- aerated concrete blocks reduced the annual collective dose by about 2%
- there was no significant radiological hazard to workers or members of the public from restored or working ash disposal sites
- there was a potential risk of increased radon exposure from buildings built from ash disposal sites. This risk could be removed by simple preventive measures at the design stage.

More recently, the UNIPEDE produced an expert group report[35] that reviewed the various features of fly ash, including the radiological properties. Their summary of the radioactivity from fly ashes around Europe is shown in Table 2.8.

UNIPEDE refers to the World Energy Conference Report, which suggests an average specific activity concentration of 200 Bq/kg. It is clear from Table 2.8 that some countries, especially those in Eastern Europe, may have difficulties with such limits. However, UK fly ashes are the lowest reported.

Puch *et al.*[36] reviewed the radioactivity of fly ash from German power stations. As in the UK, they conclude that there is only an insignificant increase in the exposure of workers, the public or within buildings resulting from the use of fly ash. Similarly, a European Commission report[37] concluded that most building materials, including those containing fly ash,

Table 2.8. Radioactivity in fly ash (Bq/kg) (UNIPEDE[35])

Reports from	Fly ash from	U-Series			Th-Series		
		Min.	Max.	Average	Min.	Max.	Average
Germany	Germany	93	137	119	96	155	121
	UK	72	105	89	3	94	68
	Australia	7	160	90	7	290	150
	Poland			350			150
Italy	Italy	130	210	170	100	190	140
Denmark	Denmark	120	210	160	66	190	120
Sweden	Sweden	150	200		150	200	
Belgium	Belgium	112	316	181	88	277	150
Spain	Spain	80	106	91	77	104	89
Germany				189			118
Czech Republic	Czech Republic	35	190	129	62	142	90

Fig. 2.7. Aerated concrete blocks made from >90% fly ash

would not present a significant risk, with the exception of natural building stone, which may represent a risk in certain circumstances.

In conclusion, radiation from coal fly ash would not normally present any significant risk to workers or the public from the coals in use in the UK and most of Europe (Fig. 2.7). However, some coals from Eastern Europe have a level of natural radiation that could give cause for concern in some fly ash-intensive applications.

UK health and safety requirements

Fly ash and FBA, in general, are not considered hazardous to health but should be handled in accordance with good occupational hygiene and safety practices. Assessment of possible exposure to constituents of the ash particles such as arsenic and chromium indicates that the content of any particular compound is so low that health effects due to exposure to the dust nuisance will occur long before any specific toxicity limit is reached. The total exposure to harmful compounds is unlikely to exceed natural background levels received from other dust sources in the environment, such as soil. The following information is that required in the UK for the Control of Substances Hazardous to Health (COSHH) Regulations 1994.

Exposure to dust

Extensive testing[38] has shown that fly ash and FBA are non-toxic and environmentally benign. Fly ash and FBA have not been assigned specific occupational exposure limits within the UK, but exposure to airborne dust may cause irritation to the eyes and the respiratory system. Within the UK, the personal exposure limits require the user to keep dust to the minimum that is reasonably practical. The COSHH Regulations 1994 require that airborne dust should not exceed $10\,mg/m^3$ in an 8 h time-weighted average (TWA) total inhalable dust. Respirable dust exposure should not exceed $5\,mg/m^3$ in an 8 h TWA. Monitor as for airborne inhalable dust, by gravimetric determination.

Contact with the skin

Current information suggests that there is no epidemiological evidence of a significant health risk associated with fly ash and FBA. When damp fly ash and FBA are moderately alkaline, prolonged skin contact with these materials may result in abrasion and irritation.

Handling precautions

Avoid creating airborne dust wherever possible. Where dust is generated then engineering control measures should be considered (water sprays) to keep the airborne dust concentration as low as is reasonably practical.

Avoid prolonged skin contact, especially where the product is dampened. Wear protective clothing, e.g. goggles, gloves, overalls and boots. Change heavily contaminated clothing as soon as possible; launder before re-use. Good housekeeping practices as well as high standards of personal hygiene should be maintained.

The use of respiratory equipment must be strictly in accordance with the manufacturer's recommendations and any statutory requirements governing its selection and use.

Fire and explosion information

There are no risks of fire or explosion as the by-products identified are non-combustible.

First-aid treatment

The following are the current UK recommendations:

- *Skin:* Wash contaminated areas of the body with soap and water as soon as is reasonably practical.
- *Eyes:* If the substance has entered the eyes then irrigate with emergency eye wash solution (if available) or clean water for up to 15 min. Obtain medical advice if any pain or redness persists.
- *Inhalation:* If inhalation of the dust causes irritation of the nose or coughing remove the patient into fresh air. Keep warm and at rest. Carefully remove any excess dust from nasal passages and rinse mouth with water until clear. If symptoms persist obtain medical advice.
- *Ingestion:* There are no known adverse affects. Wash mouth out with water and give water to drink. Do not induce vomiting.

In all cases, should exposure be excessive or symptoms develop seek medical attention.

Fly ash and FBA are composed of inorganic material with a small proportion of carbon particulate resulting from the incomplete combustion of the parent fuel, coal (Table 2.9). FBA is extracted from the combustion chamber by a hydraulic process. Fly ash is extracted from the flue gases discharged from the combustion processes by electrostatic and mechanical extraction techniques.

Handling and storage

The following are the UK practice and precautions that should be taken when handling or storing fly ash and FBA:

- *In dry form:* Keep in containers or silos or in sealed bags.
- *In conditioned, lagoon or with added water forms:* When stored in stockpiles keep exposed surfaces damp and cover small stockpiles with protective sheeting.

- *Transporting in dry form:* Transport in sealed tankers or similar units.
- *Transporting conditioned or lagoon or with added water forms:* In open vehicles with exposed surfaces protected with sheeting.
- *Classification and transport:* Fly ash and FBA are not classified as dangerous under the Classification Packaging and Labelling of dangerous Substances Regulations (Table 2.10). They are not classified dangerous for road, rail, sea or air transport.

Table 2.9. Health and safety properties of fly ash and furnace bottom ash (FBA)

Property	Range of values for UK fly ash and FBA
Particle density (specific gravity)	1·8–2·4
Solubility in water	<3%
Bulk density	1·2–1·7 g/cm^3
Alkalinity (pH)	9–12 when damp
Boiling point/boiling range	N/A
Melting point/melting range	N/A
Flash point	N/A
Flammability and autoflammability	N/A
Oxidising properties	N/A
Vapour pressure	N/A

Table 2.10. Chemical constituents of fly ash and furnace bottom ash

Component	Average % by weight
SiO_2	45–51
Al_2O_3	27–32
Fe_2O_3	7–11
CaO	1–5
MgO	1–4
K_2O	1–5
Na_2O	0·8–1·7
TiO_2	0·8–1·1
SO_3*	0·3–1·3
Cl	0·05–0·15

*Water soluble.

The figures for SiO_2 refer not to free silica but to silicon present as silicates of varying compositions.

Accidental release measures

Prevent entry into drains and watercourses. Large spills of dry material should be removed by a vacuum system. Conditioned (dampened) material should be removed by mechanical means where possible and be recycled or disposed of in a licensed site.

Applying a fine water spray can reduce the potential for dust blow.

Mobility, persistence and degradability, bioaccumulation potential and aquatic toxicity

Fly ash/FBA has no known ecotoxic effects in the existing patterns of production, handling, storage and use.

Disposal

Fly ash and FBA are classed as 'Controlled Wastes' in the UK and have no special requirements for their disposal at appropriately licensed facilities. They are included in the European Waste Catalogue (Code No. 10 01 02) but are not hazardous materials as determined by EC Hazardous Waste List (Directive 94/904/EC). They are also 'Green List' materials for trans-frontier shipment. The CAS number is 68131-74-8. The ACX number is X1014150-5.

Summary of properties of fly ash

UK fly ash is a fine, silt-like material with surprisingly consistent chemical and physical properties. It has the potential for use in many applications with beneficial results. It can be used for many different applications, e.g. as a fill material, as a fine aggregate for grouts, and in concrete as both an aggregate and a cementitious component. In environmentally sensitive areas simple, common-sense precautions are required to prevent any risk of contamination, e.g. boron-tolerant planting or protection of fill.

References

1. Parrott L. Environmental report for the UK concrete industry 1994 to 1998. *CIA/DETR project – Defining and improving environmental performance in the concrete industry*. British Cement Association, Crowthorne, January 2000.
2. Parrott L. Effects of ground granulated blastfurnace slag and pulverised-fuel ash upon the environmental impacts of concrete. *CIA/DETR project – Defining and improving environmental performance in the concrete industry*. British Cement Association, Crowthorne, January 2000.

3. Meij R, Schaftenaar H. *Hydrology and chemistry of pulverised fuel ash in a lysimeter – the translation of the results of the Dutch column leaching test into field conditions*. Kema, Arnhem, 1994.
Fruchter JS, Rai D, Zachara JM. Identification of solubility-controlling solid phases in a large fly ash field lysimeter. 1990; 1173–1179.
4. Brown J, Ray NJ, Ball M. The disposal of pulverised fuel ash in water supply catchment areas. *Water Research* 1976; **10**: 1115–1121.
5. Morley Davies W, Gillham EWF, Simpson DT. An investigation into farming on land restored with fly ash. *Journal of the British Grassland Society* 1971; **26**: 25–30.
6. Brown J, Ray NJ. The handling and disposal of coal ash in the CEGB in relation to the aqueous environment. *Water Science and Technology* 1983; **15**: 11–24.
7. Mattigod SV, Eary LE, Dhanpat R. Geochemical factors controlling the mobilisation of inorganic constituents from fossil fuel combustion residues 1, review of major elements. *Journal of Environmental Quality* 1990; **19**: 188–201.
8. Eary LE, Rai D, Mattigod SV *et al*. Geotechnical factors controlling the mobilisation of inorganic constituents from fossil fuel combustion residues: Part II – Review of minor elements. *Journal of Environmental Quality* 1990; **19**: 202–214.
9. Gillham EWF. *Trace elements in soils and ashes – a comparative view*, Vols 1 and 2. CEGB, London, 1980.
10. Commission of the European Committees. *Leaching and soil/groundwater transport of contaminates from coal combustion residues.* Report EVR 14054EN, 1992.
11. Various European standards are under preparation by TC292 – *Characterisation of waste*, e.g. the prEN12457 series.
12. Proposed European standard. *Characterisation of waste – Compliance leaching test for monolithic material*, ~2004.
13. Ray NJ, Ball M. The percolation of water through pulverised fuel ash. CEGB Internal Document, London, 1972.
Ray NJ, Ball M. Percolation of water through pulverised fuel ash: Part 3 – Experiments under natural weathering conditions. CEGB Internal Document, London, 1972.
Ray NJ, Ball M. Contributions of fly ash lagoon effluents to the chemical constitution of the River Trent. CEGB Internal Document, London, 1973.
14. DIN 38414 Part 4. *German standard methods for the examination of water, waste water and sludge. Sludge and sediments (group S). Determination of leachability by water.*
15. Scott PE, Baldwin G, Sopp C *et al*. *Leaching trials on materials proposed as infill for Combe Down stone mines*. AEA Technology Report AEA/CS/18303036/019, 1994.
16. Young PJ, Wilson DC. *Testing of hazardous waste to assess their suitability for landfill disposal*. AERE Report R10737, 1987.
17. Baldwin G, Addis R, Clark J, Rosevear A. *Use of industrial by-products in road construction – water quality effects*. CIRIA Report 167, 1997.

18. Wild SR, Jones JC. Polynuclear aromatic hydrocarbons in the United Kingdom environment: a preliminary source inventory and budget. *Environmental Pollution* 1995; **88**: 91–108.
19. Sharkey AG, Schultz JL, White C, Lett R. *Analysis of polycyclic organic material in coal, coal ash, fly ash and emission samples*. Report No. EPA-600/2-76-075. United States Environmental Protection Agency, 1976.
20. Wright RD. *Polycyclic aromatic hydrocarbon compounds in pulverised fuel ash from CEGB power stations*. Report No. TPRD/L/2924/N85. CEGB, London, 1986.
21. Zenon Environmental Inc. Characterisation of organic contaminants in ash samples from pulverised coal-fired power generating stations. *Environment Canada* 1987.
22. National Rivers Authority. *Protocol for a leaching test to assess the leaching potential for soils from contaminated sites*. NRA R&D Note 181. NRA, DETR, London.
23. Janssen F, Kanij J. The trace analysis of polycyclic aromatic hydrocarbons (PAH) adsorbed on coal fly ash. *Mikrochimica Acta (Wien)* 1984; **1**: 481–486.
24. Natusch DFS. *Formation and transformation of polycyclic organic matter from coal combustion*. Progress Report, Colorado State University, 1979.
25. Harrison FL, Bishop DJ, Mallon BJ. *The kinds and quantities of organic combustion products in solid and liquid wastes from a coal-fired power station*. Lawrence Livermore Laboratory, 1983.
26. Junk GA, Richard JJ, Avery MJ. *Organic compounds in effluents related to coal combustion*. Pre-prints of papers. American Chemical Society, Division of Fuel Chemistry, 1985; **30**(2): 171–178 ACS.
27. Mamantov G, Wehry EL. *Chemical interactions of polycyclic organic compounds with coal fly ash and related solid surfaces*. Report No. DOE/ER/60552–2. US Department of Energy, Washington, DC, 1989.
28. Mamantov G, Wehry EL. *Sorption and chemical transformation of PAHs on coal fly ash*. Report No. DOE/PC/19306-14. US Department of Energy, Washington, DC, 1995.
29. Freedman AN. *The analysis of power station fly ash for the presence of polychlorinated dibenzo-p-dioxins*. CEGB, London, 1988.
30. See UKQAA website on www.UKQAA.org.uk for information about the proposed code of practice.
31. European Commission. *Development of a leaching method for the determination of the environmental quality of concrete*. EUR 17869 EN, 1997. ISSN 1018–5593.
32. Lloyd T, Wilson I, Concrete in contact with drinking water. *Society of Chemical Industries seminar*, 25th May 2001, London, UK.
33. Woolley GR, Simpson DT, Quick W, Graham J. *Ashes to assets*. PowerGen UK plc, 2000. ISBN 0-9516457-06.
34. Green BMR. *Radiological significance of the utilisation and disposal of coal ash from power stations*. CEGB, National Radiological Protection Board, January 1986.
35. UNIPEDE. *Coal Ash Reference Report*. Thermal Generation Study Committee Report, 20.05 THERRES, 1997.

36. Pugh KH, Keller G, vom Berg W. *Radioactivity of combustion residues from coal fired power stations*. VGB Technical Association of Large Power Plant Operators, Essen, 1996.
37. European Commission. *Radiological protection principles concerning the natural radioactivity of building materials*. Radiation Protection 112, Directorate-General, Environment, Nuclear Safety and Civil Protection, 1999.
38. Meldrum M, Maidment S, Gillies C. *Pulverised fuel ash: criteria document for an occupational exposure limit*, HSE document EH 65/2.

Chapter 3

Using fly ash in concrete

The history of concrete and the use of pozzolanic materials from Roman times to date

Fly ash is a pozzolana, that is a material that reacts with lime to form a hardened mass. By mixing a red volcanic powder, found at Pozzuoli near Naples, with lime the Romans discovered in 75 BC that a form of concrete could be created.[1] The word concrete comes from the Latin *concretus* which means grown together or compounded. Similarly, Rivero-Villarreal and Cabrera[2] reported on the properties of a lightweight concrete developed around the same time in the ancient culture of Totonacas near the modern city of Veracruz, Mexico. Pozzolanic cements were first used at Pompeii in 55 BC. For many years, these concretes were only used as an infill material between walls. Gradually, concrete replaced the brick and stone as in the arches of the Colosseum and the Pantheon, constructed in AD 115. The dome of the Pantheon is an example of pozzolanic lightweight concrete, which is over 50 m in diameter and made with a lightweight aggregate (pumice), with an air-entraining agent (animal blood) and a pozzolanic material (volcanic ash) (Fig. 3.1).

With the decline of the Roman Empire concrete disappeared until Norman times. One of the first Norman structures was Reading Abbey, built in AD 1121. Concrete was not used other than for foundations until John Smeaton began experimenting in 1756. He used Aberthaw Blue Lias and a pozzolana from Civitavecchia, near Rome, to build the Eddystone lighthouse. This was later moved to Plymouth Hoe where it still stands today. In 1824, Joseph Aspdin obtained his patent for 'Portland cement'. Owing to the more rapid gain in strengths of Portland cement, pozzolanic or 'Roman cements' fell out of use. However, one of the hydration by-products of Portland cement is lime. If pozzolanic materials are combined with Portland cement the lime produced from cement hydration reacts with the pozzolana. It is this reaction that led to the usage of fly ash within concrete in later years.

Fig. 3.1. The Pantheon in Rome

The first reference to the idea of using coal fly ash within modern-day concrete was by McMillan and Powers[3] in 1934 and research in the USA[4] indicated that fly ash had a role in concrete in 1935. Later research, also in the USA,[5] reported that fly ash was a possible artificial pozzolana. Trial applications and continuing research promoted the idea that introduction of a proportion of fly ash as replacement of cement would limit shrinkage cracking in mass concrete by reducing internal hydration temperatures.

The introduction of modern steam raising plant in the UK, particularly after World War II, gave access to fly ash and the late 1940s saw research into the use of the material. In particular, the example of using fly ash in mass concrete dams was considered and, following research at the University of Glasgow,[6] the practice was adopted for construction of the Lednock,[7] Clatworthy and Lubreoch dams. These dams formed part of the Scottish Hydro-Electric Board's Breadalbane scheme. Lednock involved some 82,000 cubic yards (62,500 m^3) of concrete, saving some 3000 tonnes of Portland cement. The control criteria were somewhat crude, using the colour and the gritty feel of the ash. It was found that the variability of the fly ash both in fineness and in carbon content was problematical. The power station supplying the ash used some 20 differing coal sources during the period of construction and carbon content was not monitored or controlled in any manner. However, the subsequent durability of the structure has been excellent.

There followed, in the period 1954–1958, examples of the use of fly ash as cement replacement in structural concrete at the Fleet Telephone Exchange, Newman Spinney Power Station[8] and the High Marnham substation.[9] By the mid 1970s fly ash was regularly being used in concrete as an addition at the concrete plant within many power company structures and some notable public works[10] being constructed. Such usage was always on a basis of close monitoring by the site and within large construction projects. In the UK, fly ash from coal combustion became known as pulverised fuel ash (PFA, or fly ash) around this time to differentiate it from ashes derived from other processes.

Although the use of fly ash or PFA in concrete was accepted by British Standards it was not until 1965, when the first edition of BS 3892,[11] was published that there was a standard for the PFA for use in concrete. PFA was treated as a fine aggregate with three classes of fineness based on the specific surface area. During this period acceptance in the routine readymixed concrete supply market was not being achieved. During the 1970s readymixed concrete suppliers were producing ever more technically demanding concretes of higher strength and lower water/cement (W/C) ratios. It was perceived that the variability in quality and the supply problems of fly ash when taken directly from the power station were unacceptable.

Precast concrete manufacturers are somewhat more tolerant to some variability as they have control of the whole process, being able to control the curing methods and time before dispatching the finished concrete element. In addition, they tend to have close working relationships, and have developed individual specifications, with the power stations. As a final resort they can adjust their process to compensate for any fly ash variability.

Variability in the fly ash fineness, with all other factors being constant, leads to variation in the water content and strengths of the resulting concrete for a given ash source. Variations in loss on ignition (LOI) lead to colour variations and difficulties when trying to entrain air for frost-resistant concrete. The variability stems from the limitations of the power production process. Power can only be produced when it is needed, as electricity storage on any scale is highly complex. Therefore, many coal-fired power stations only operate when there is a high demand. Furnaces are started and stopped, which leads to a variable quality of fly ash. In the summer months electricity consumption reduces and only the 'base load' power stations are able to provide fly ash of consistent fineness and LOI.

In the UK one solution was found to many of the fineness variability problems when in 1975 Pozzolan Ltd introduced the concept of supplying controlled fineness material. Controlling the fly ash/PFA to a tightly controlled fineness involved either classifying the ash, to remove the coarse fractions, or selection of the finer material by continual monitoring of the ash production. In general, classification enhances the pozzolanicity whilst reducing the water demand. An Agrément Board Certificate[12] was obtained for classified fly ash, or PFA, in 1975. These changes were reflected within BS 3892[13] in 1982 with the various parts of the standard indicating the uses and quality of PFA. Classified PFA to BS 3892 Part 1 was accepted as counting fully towards the cement content of a mix, whereas 'run of station' ashes were at the discretion of the site engineer. The latter were usually considered as inert fillers and are covered by BS 3892: Part 2, 1984.

In 1985 two British Standards were published for cements containing PFA:

- BS 6588 for Portland PFA cements permitted an ash level between 15% and 35% by mass of cement
- BS 6610 for pozzolanic PFA cement permitted an ash level of 35% and 50% by mass of cement.

Before 1985, interground Portland fly ash cement had been produced by Blue Circle in the North of England, under an Agrément Certificate.

Classified PFA was increasingly accepted for use within concrete on both technical and economic grounds. Currently, the use of classified PFA is widespread within the UK readymixed and precast concrete industries. Some 25% of the readymixed concrete produced in the UK contains a binder that consists of, typically, 30% PFA and 70% Portland cement. Currently, some 500,000 tonnes per annum of classified PFA is used in readymixed and precast concrete. With European harmonisation, a new standard for fly ash, BS EN 450[14] 1995, was introduced. With the exception of the UK and Ireland, no other European countries routinely classify fly ash for use in concrete. EN 450 reflects this differing approach and allows a wider range of fineness for use in concrete than BS 3892 Part 1.[15] The enabling standard for EN 450 fly ash, EN 206,[16] has taken a number of years to finalise and consequently the use of EN 450 fly ash is restricted. Within EN 206 the primary method of use only allows EN 450 fly ash to be partially counted towards the cement content of the mix using the 'k' value approach developed by Smith.[17] A maximum of 25% of the combination can be counted as cement within EN 206. An alternative route permitted within EN 206 is the equivalent concrete performance concept where it is required to show equal performance with a reference concrete. One approach by Dhir *et al.*[18] proposes the use of equal strength class concrete to give equal durability. Changes to either binder content, water content or both are proposed as ways of achieving equivalence. This method accepts the better performance of finer fly ash. In order to allow UK practice to continue, classified PFA to BS 3892 Part 1 will remain permitted under BS 8500[19] for the foreseeable future.

British and European standards for fly ash in concrete

Additions of fly ash during concrete manufacture

An addition is defined in BS EN 206 as a finely divided inorganic material used in concrete to improve certain properties or to achieve special properties. There are two types of addition:

- type I: these are nearly inert additions
- type II: these are pozzolanic or latent hydraulic additions

Many additions of both types are available:

- PFA/fly ash (BS 3892 and BS EN 450): type II
- ground granulated blastfurnace slag (GGBS): type II
- filler aggregates (prEN 12620[20]): type I
- pigments for building materials (BS EN 12878[21]): type I
- metakaolin and silica fume (prEN 13263[22]): type II.

These may be used singly or in combination. Combinations of PFA to BS 3892: Part 1, with Portland cement to BS 12 (or BS EN 197-1[23] CEM I), count fully towards the cement content and W/C ratio in concrete provided that they have satisfied the equivalence testing procedures set out in the annex to BS 3892 (also as an annex within BS 8500[24]).

BS 3892 Parts 1, 2 and 3

BS 3892 Part 1[15] covers PFA for use in concrete. This is a type II addition and counts fully towards the cement content of the mix. The standard imposes stringent quality-control requirements on the supplier and the fly ash must be obtained from the flue gases of power stations burning bituminous or hard coal. The PFA must be processed to meet specific requirements on fineness, LOI, strength factor and water demand. Strength factor is the ratio of strength of a 30% fly ash mix to a Portland cement-only mix. The mixes are mortars compacted into prisms produced to the requirements of BS EN 196-1,[25] except that the water content of the fly ash mix is adjusted such that it has the same flow as the control, which is a plain CEM I-only mortar.

BS 3892 Part 2[26] covers type I addition of PFA to concrete. These PFAs are not processed and fineness and LOI limits are considerably less restrictive than BS 3892 Part 1 and encompass all fly ash produced. There are no limits on water requirement. However, as a type I addition such PFA cannot be counted towards the cement content and is considered to be an inert filler.

BS 3892 Part 3[27] covers PFA for use in cementitious grouts (see Chapter 6).

BS EN 450

BS EN 450[28] covers fly ash produced from the burning of hard coal in power stations, but has differing requirements from BS 3892 Part 1 PFA for fineness and LOI. It introduces the concept of the activity index (AI). The AI is the ratio of strengths of mortar prisms for a 25% fly ash plus CEM I mortar mixture against a CEM I only mortar at fixed W/C ratio. Testing is carried out to the requirements of BS EN 196-1. Minimum values for AI are required at 28 and 90 days. The major difference from BS 3892 strength factor requirements is that the mortar is prepared at a fixed W/C ratio. Any water-reducing properties of the fly ash are effectively ignored.

BS EN 450 is based around the supplier providing a fly ash which does not need processing but does require the supplier to demonstrate consistency. Processing is allowed if the supplier wishes to improve the properties and/or consistency of the ash. The quality-control requirements are not as onerous as BS 3829 Part 1 fly ash but more severe than in BS 3892 Part 2. The requirements are summarised in Table 3.1.

Properties of PFA to BS 3892 Part 1 and fly ash to BS EN 450

In 1999 NUSTONE carried out a programme of testing[29] of UK fly ash from five power stations using the test methods described in BS 3892 Part 1 and BS EN 450. This programme was designed to assess the performance and variability of UK ashes when tested to the two standards.

Figure 3.2 shows the variation found in fly ash both before (EN 450 fly ash) and after classification (BS 3892 Part 1 PFA), showing the reduction in both fineness and variation possible by classification.

All samples were tested using both standards' testing regimes. It was found the BS 3892 Part 1 requirements were more onerous as a fixed workability of the mortars tested is required. Figure 3.3 shows the strength factor results and the high degree of scatter associated with taking a ratio of two strengths, e.g. the strength of the fly ash prism divided by the Portland cement prism. When one takes an overview of the ashes from all the sources and only uses the fly ash prism strength a clearer picture is seen, as in Fig. 3.4.

Clearly, the finer the ash the greater the reactivity at 28 days. Further analysis of the data confirms that this is primarily due to the water-reducing properties of the finer fly ash, although there is an element of source dependency (Fig. 3.5).

One argument for classifying fly ash has been the reduction in variability that is achieved. Although classification improves the mean strength performance of fly ash and reduces the variation in fineness, the inherent strength variability of unclassified fly ash was not found to be significantly different from classified ash, as shown in Table 3.2 and Fig. 3.6.

Table 3.1. Summary of the requirements of the various UK fly ash/pulverised fuel ash (PFA) standards

Attribute	Requirements		
Standard	PFA BS 3892 Part 1	PFA BS 3892 Part 2	BS EN 450 fly ash
Particle density	⩾2000 kg/m^3	N/A	±150 kg/m^3 of declared value
Fineness	⩽12% retained 45 μm sieve	⩽60% retained 45 μm sieve*	⩽40% retained 45 μm sieve Must be within ±10% of declared value
Soundness	⩽10 mm 30% fly ash + 70% PC (BS 12 42.5)	N/R	⩽10 mm 50% fly ash + 50% CEM I 42.5
Sulfur: maximum present as SO_3	⩽2·0%*	⩽2·5%*	⩽3·0%
Chloride	⩽0·10%	N/A	⩽0·10%
Calcium oxide	Expressed as total CaO ⩽10%	N/R	Expressed as free CaO ⩽1·0% or ⩽2·5% if soundness satisfactory
Loss on ignition	⩽7·0%*	⩽12·0%*	⩽7·0%†
Moisture content	⩽0·5%	⩽0·5% unless conditioned ash used	Must be dry
Water requirement	⩽95% of PC 30% fly ash + 70% PC (BS 12 42.5)	N/A	N/A
Activity index: ref. EN 450 – EN 196-1	N/A	N/A	⩾75% at 28 days ⩾85% at 90 days 25% fly ash + 75% CEM I 42.5
Strength factor: ref. BS 3892 Part 1 Annex F	⩾0·80 at 28 days	N/A	N/A

*Absolute limits. Other values are autocontrol limits.
†Permitted on a national basis only.

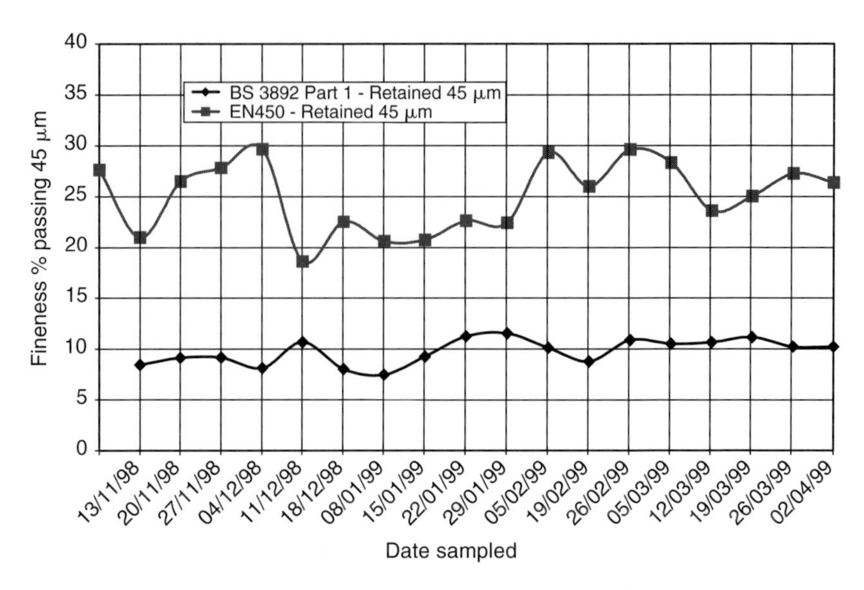

Fig. 3.2. Typical variation in fineness found from a single UK power station (Source B – EN 450 and BS 3892 Part 1 – fineness vs time)

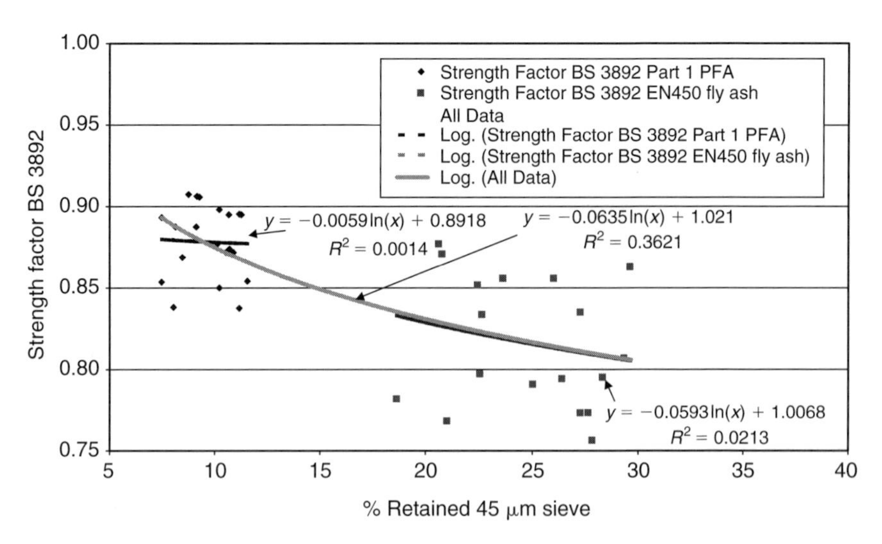

Fig. 3.3. Effect of fineness on strength factor (BS 3892 testing regime – strength factor vs fineness source B)

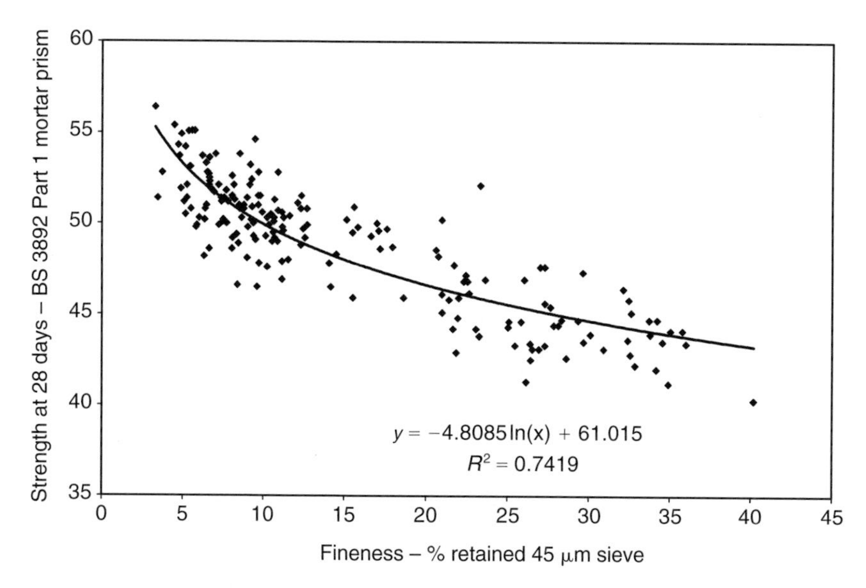

Fig. 3.4. Effect of fineness on mortar prism strength (BS 3892 Part 1 testing regime; all sources)

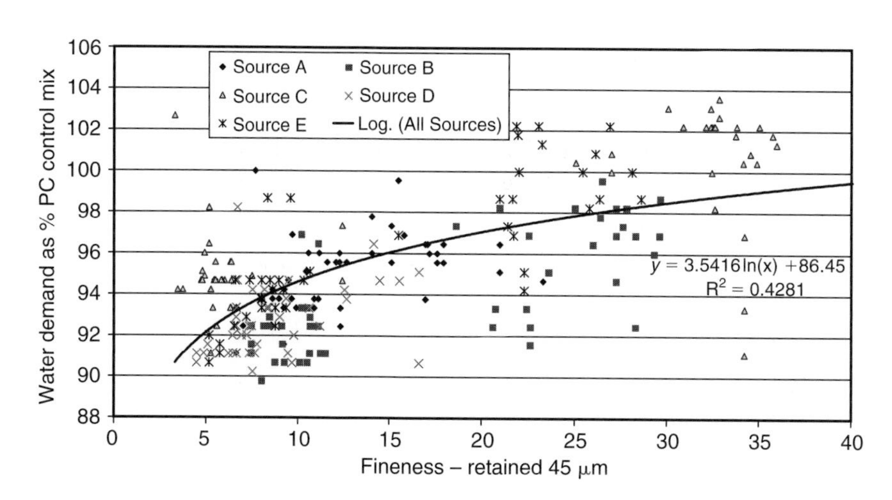

Fig. 3.5. Effect on water demand of fly ash fineness (BS 3892 Part 1 testing regime)

Table 3.2. Confidence limits for mean fineness and standard deviations of strengths

	SD of strengths (MPa)		95% confidence limits for mean fineness (retained 45 μm sieve) (Based on 20 results)
	BS 3892 Part 1 – BS 3892 testing regime	EN 450 – BS 3892 testing regime	
Source A	1·528	1·189	±1·517
Source B	1·609	1·741	±1·505
Source C	2·270	1·718	±1·366
Source D	1·568	1·977	±1·365
Source E	2·076	1·655	±1·898
Overall	1·810	1·656	±1·530

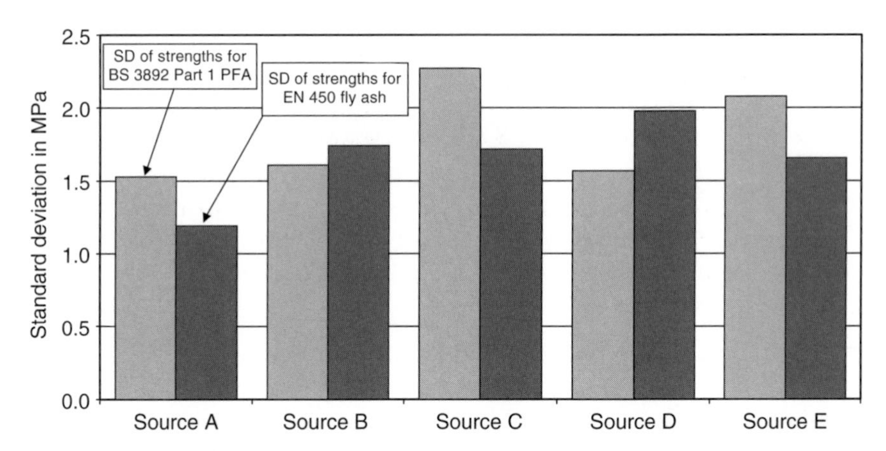

Fig. 3.6. Comparison of standard deviations for fly ashes (BS 3892 testing regimes)

The following conclusions were drawn from the data.

- The fineness of fly ash has a significant effect on the water demand and therefore the strength performance of the mortar when tested to EN 196-1. Finer fly ashes give greater water reductions and improved strengths.
- The source of the fly ash has a significance effect on the strength performance. This effect is reduced when the ash is classified to comply with BS 3892 Part 1, but the reason for this is unclear. It may result from the lack of information about the particle size distribution given by measuring the percentage passing the 45 μm sieve.
- The BS 3892 testing regimes, which rely on testing at a fixed workability, are more sensitive to changes in the properties of the fly ash

than the equivalent EN 450 test. The variability in strength when using EN 450 fly ash is not statistically significantly different from that found with BS 3892 Part 1 fly ash when tested in mortar prisms.
- The fineness when measured as the percentage retained on the 45 μm sieve only gives an indication of the particle size distribution properties of a fly ash.

Cements containing fly ash

BS EN 197-1: 2000[23] contains a variety of cements, of which a number contain fly ash (Table 3.3).

There are a few restrictions of the types of fly ash permitted:

- The reactive silica content shall not be < 25·0% by mass.
- The loss on ignition for the fly ash shall not exceed 5·0%; 7·0% LOI is permitted where allowed in the appropriate standards applicable in the place of use, e.g. the UK.
- Reactive CaO shall be < 10% and free CaO less than 1·0%. Free CaO contents of >1·0% and <2·5% are permitted subject to a requirement on expansion testing (soundness). It shall not exceed 10 mm when tested in accordance with EN 196-3 using a mixture of 30% fly ash and 70% by mass of a CEM I cement conforming to EN 197-1.
- The fly ash shall be proven to be pozzolanic when tested in accordance with EN 196-5.[30]

Compliance with the various requirements for all common cements, e.g. soundness, LOI and sulfate content, assures the user that unsuitable fly ashes are not used.

In addition to the cements permitted under the European standard, pozzolanic pulverised fuel ash cement is permitted in the UK as BS 6610. This was revised in 1991, with a minor revision in 1996, to bring these cements in line with BS EN 197-1. The main features of BS 6610 are summarised in Table 3.4. These cement standards form the basis of the equivalence rules, which govern the use of fly ash in concrete.

Using fly ash within concrete: BS 5328 and EN 206

BS 5328[31] Parts 1–4 allow the use of BS 3892 Part 1 PFA as a mixer addition and the factory-made cements, BS EN 197-1 and BS 6610[32], over the full range of concretes. BS 3892 Part 1 PFA additions and factory-made cements all count fully towards the cement content of the mixes. Where increased cement contents are felt necessary owing to durability requirements these are given as separate figures within the various tables. These depend on the proportion of fly ash being used and the specific durability

Table 3.3. Cement types containing fly ash (extract from BS EN 197-1, Table 1)

<table>
<tr><th rowspan="3">Main cement type</th><th rowspan="3">Designation</th><th rowspan="3">Notation</th><th colspan="3">Constituents
Proportion by mass (%) based on the sum of the main and minor constituents</th><th rowspan="3">Minor additional constituents</th></tr>
<tr><th rowspan="2">Clinker</th><th colspan="2">Fly ash</th></tr>
<tr><th>Siliceous</th><th>Calcareous</th></tr>
<tr><td>CEM II</td><td>Portland fly ash cement</td><td>CEM II/A-V</td><td>80–94</td><td>6–20</td><td></td><td>0–5</td></tr>
<tr><td></td><td></td><td>CEM II/B-V</td><td>65–79</td><td>21–35</td><td></td><td>0–5</td></tr>
<tr><td></td><td></td><td>CEM II/A-W</td><td>80–94</td><td></td><td>6–20</td><td>0–5</td></tr>
<tr><td></td><td></td><td>CEM II/B-W</td><td>65–79</td><td></td><td>21–35</td><td>0–5</td></tr>
<tr><td>CEM IV</td><td>Pozzolanic cement</td><td>CEM IV/A</td><td>65–89</td><td colspan="2">11–35*</td><td>0–5</td></tr>
<tr><td></td><td></td><td>CEM IV/B</td><td>45–64</td><td colspan="2">36–55*</td><td>0–5</td></tr>
</table>

*May also be natural pozzolana.

Table 3.4. Summary of BS 6610 requirements

BS number:	BS 6610: Pozzolanic PFA cement 1995
Fly ash	Must meet the requirements of BS 3892 Part 1 PFA, except for fineness if interground
Fly ash level*	36–55%
Minor additional constituents	0–5%
Compressive strength (EN 196-1 mortar prism test)	7 days: 12 N/mm^2 28 days: 22.5 N/mm^2

*As % nucleus, i.e. clinker plus fly ash.

needed. BS 3892 Part 2 PFA is also permitted but does not count towards the cement contents in any way. However, the standard does permit the use of 'non-standard' materials and requires 'satisfactory data on their suitability and assurance of quality control'. Using this route, many structures have been built using a wide range of fly ash grades, which have proven to be durable in the fullness of time.

BS EN 206 has a far more complex approach to the utilisation of fly ash. BS EN 206 places little restriction on the use of additions, simply stating that additions of type I and type II may be used in concrete in quantities as used in the 'initial tests'. Initial tests are defined in an annex of BS EN 206 as those required for demonstrating that a mix satisfies all specified requirements for the fresh and hardened concrete. These initial tests may be from laboratory work or from long-term experience.

The situation becomes more complex when additions are taken into account as part of the total cementitious content and when calculating the W/C ratio. BS EN 206 contains specific rules for fly ash to BS EN 450 and these rules may be applied anywhere. It also permits the use of other rules, if the suitability of the rules is established.

The specific rules in EN 206 are based on the '*k*-value' concept:

- the term water/cement ratio is replaced by water/(cement + $k \times$ addition).
- The minimum cement content (MCC) can be reduced by $k \times (\text{MCC} - 200)$ kg/m^3. However, the amount of cement plus fly ash must never fall below the MCC value.
- Fly ash to BS EN 450 has a *k*-value of 0.2 or 0.4, depending on the strength class of the Portland cement with which it is used. The *k*-value does not vary with the quantity of ash being used. Up to a maximum of 25% fly ash by mass of the (cement + ash) is allowed to be counted as cementitious. Any additional ash within the mix is effectively a type I addition, which is assumed to act as an inert filler.

Other values of *k* or other *k*-value concepts may be used if their suitability is established. One way of establishing suitability is via a National Standard valid in the place of use of the concrete (Table 3.5). BS 8500[24] uses this option by including rules by which PFA to BS 3892 Part 1 may count fully towards the cement content and water cement ratio. These rules in BS 8500 are available for all concrete conforming to EN 206 which is to be used in the UK, but do not extend to concrete used in other countries unless permitted by a National Standard.

BS EN 206 also contains an 'Equivalent concrete performance concept' that may be applied to a combination of any specified cement with any specified addition if the suitability has been established. The test methods necessary for its implementation are not standardised. The informative Annex C of EN 206 places limits on the application of the concept.

The rules in BS 8500 are specific to both fly ash to EN 450 and PFA to BS 3892 Part 1. They are based on the testing of the combination of the addition and the cement that are from specified sources. The procedure is detailed in an annex of BS 8500 and determines permitted proportions of fly ash for a combination that can count fully towards cement content and

Table 3.5. Summary of the ways in which fly ash combinations may be used in concrete conforming to BS EN 206

Addition	Standard	Type	Permitted by	Route to count towards cement content and W/C ratio
Fly ash	BS EN 450	II	BS EN 206 as a type II addition	BS EN 206 *k*-value, proven equivalent concrete performance or compliance with an annex in BS 8500.
Pulverised fuel ash	BS 3892 Part 1	II	BS 8500 as a type II addition	Counts fully in BS 8500 when shown to be in compliance with an annex in BS 8500.
Pulverised fuel ash	BS 3892 Part 2	I	BS 8500 as a type I addition	Does not count towards the cement content unless the equivalent concrete performance route is taken.

The equivalent concrete performance concept in BS EN 206 provides an additional route for any type II addition effectively to count fully towards the cement content and water/cement (W/C) ratio (by permitting amendments to the recommended values). To take advantage of this route the manufacturing source and characteristics of both the addition and the cement with which it is used must be clearly defined and documented.

W/C ratio. The strength class of the combination is also determined. Combinations of Portland cement with BS EN 450 fly ash, which satisfy the conformity procedure in the annex, may also count fully towards the cement content and W/C ratio. Further restrictions on allowable proportions are found in BS 8500, Table of Durability Requirements.

Properties of fly ash in concrete

The beneficial effects of using fly ash mainly result from the low permeability of the resulting concrete. The low permeability is a result of the particle shape, the fineness, the chemistry and the pozzolanic reactivity of the fly ash.

The pozzolanic reaction and concrete

Fly ash from a coal-fired power station is a pozzolanic material that has a rounded particle shape. The combination of these properties gives fly ash its unique abilities of improving long-term strength gain, durability and reduced shrinkage, etc.

A pozzolana is a natural or artificial material containing silica in a reactive form. By themselves, these materials have little or no cementitious value. However, in a finely divided form and in the presence of moisture they will chemically react with calcium hydroxide (lime) to form compounds with cementing properties. It is important that pozzolanas are finely divided in order to expose a sufficient surface area to the solution of calcium hydroxide for the reaction to proceed at any detectable rate. Examples of pozzolanas are volcanic ash, pumice, opaline shales, burnt clay and fly ash. The silica in a pozzolana must be amorphous to be reactive.

When coal burns in a power station furnace at temperatures of around 1400°C the incombustible materials coalesce to form spherical glassy droplets containing silica (SiO_2), alumina (Al_2O_3), iron oxide (Fe_2O_3) and other minor constituents. The melt is of high viscosity and when the fly ash emerges from the flame and cools it remains frozen in a glassy, amorphous form. A typical UK fly ash contains 80% glassy material, with the remainder consisting mainly of silica (quartz), mullite, iron oxide and residual carbon.

When fly ash is added to concrete the pozzolanic reaction occurs between the silica glass (SiO_2) and the calcium hydroxide [$Ca(OH)_2$] or lime, which is a by-product of the hydration of Portland cement. The hydration products produced replace the partially soluble calcium hydroxide filling the interstitial pores, reducing the permeability of the matrix. Roy[33] states, 'the reaction products are highly complex involving phase solubility, synergetic accelerating and retarding effects of multi-phase, multi-particle materials and the surface effects at the solid liquid

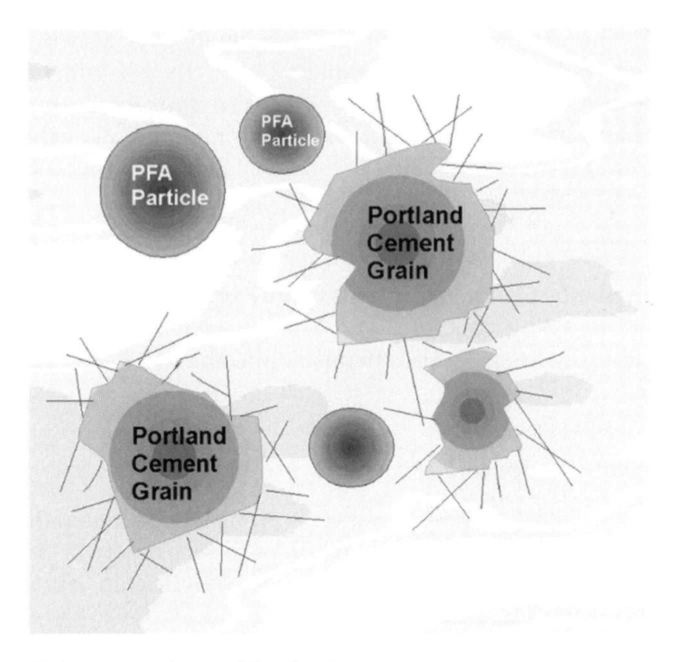

Fig. 3.7. Hydration products of Portland cement

interface'. The reaction products formed differ from the products found in Portland cement-only concretes. However, a very much finer pore structure is produced with time, presuming that there is access to water to maintain the hydration process. Dhir[34] also demonstrated that the addition of fly ash improves the dispersion of the Portland cement particles, improving their reactivity. The greater dispersion exposes a greater surface area to the hydration reaction. These processes give fly ash concrete its low permeability.

Figure 3.7 shows that when water is added to Portland cement hydration products form, locking the matrix of cement and aggregates particles together in a solid mass. $Ca(OH)_2$ (hydrated lime) is produced by the reaction, which partially goes into solution. Owing to its limited solubility particles of lime form within interstitial spaces in the matrix (Fig. 3.8). Hydrated lime is physically a weak material, so it contributes little to strength. With a continuing supply of moisture, the lime dissolves in the pore solution and reacts with the particles of the fly ash, producing further hydration products. These form a particularly fine pore structure that occupies the spaces between the various particles (Fig. 3.9). This pozzolanic reaction takes place between the glass phase of fly ash and calcium hydroxide produced during hydration of the cement largely as given below:

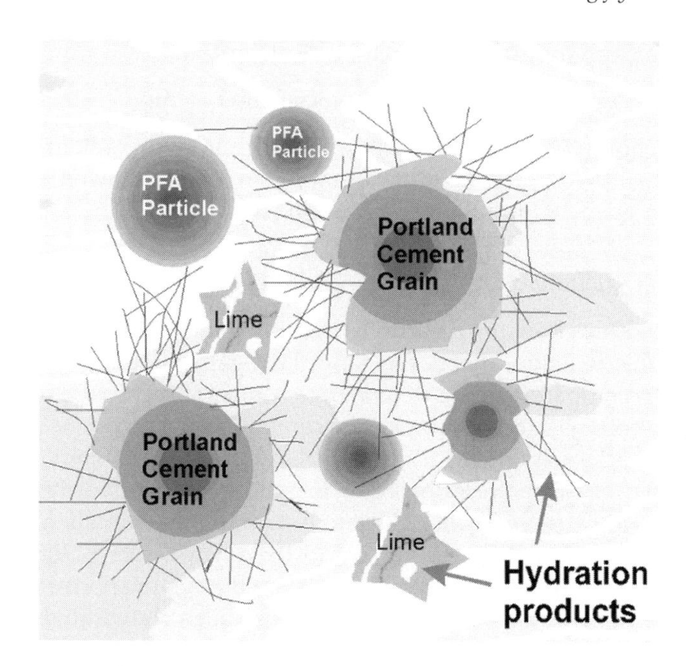

Fig. 3.8. Lime is formed as a by-product of hydration

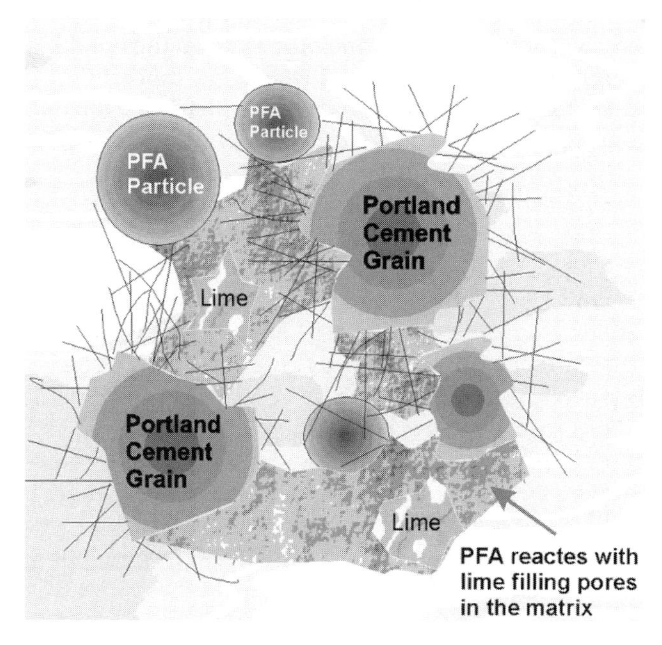

Fig. 3.9. The pozzolanic reaction products fill the interstitial spaces

Calcium hydroxide + silica = Tricalcium silicate + water

$$3Ca(OH)_2 + SiO_2 = 3CaO.SiO_2 + 3H_2O \quad (3.1)$$

This reaction requires movement of the calcium hydroxide to the reactive surface of the fly ash. It follows that if concrete is allowed to dry out then this movement, and the reaction, will cease.

Depletion of calcium hydroxide

Figure 3.10 shows how the calcium hydroxide (lime) is depleted with time and how this reaction affects the long-term gain in strength of fly ash concrete (A) compared with a PC concrete control (B). The reaction, which takes place both within the pores of the cement paste and on the surface of fly ash particles, produces calcium silicate and aluminate hydrates. Despite the pozzolanic reaction reducing the available hydrated lime in the pore solution, there is still sufficient remaining to maintain a high pH.

Research has shown that pore size in fly ash concrete is significantly smaller than the equivalent PC concretes, even though the porosity, a measure of pore volume, may be greater. It has also been shown that pore size in fly ash concrete continues to reduce with time beyond that experienced with an equivalent PC concrete.[36] It follows that a well-compacted fly ash concrete has a lower permeability. Studies of the oxygen permeability of fly ash concrete found it to be lower than a similar PC concrete, even after only 1 day's curing.[37] This reaction and the physical outcome reduce porosity in fly ash concrete with time and increase the bond between the paste and particles.[38] This is illustrated well in Fig. 3.11.

Some pozzolanic reactions begin during the first 24 h. However, the contribution to strength of fly ash is normally far less than that of PC. Thus, for

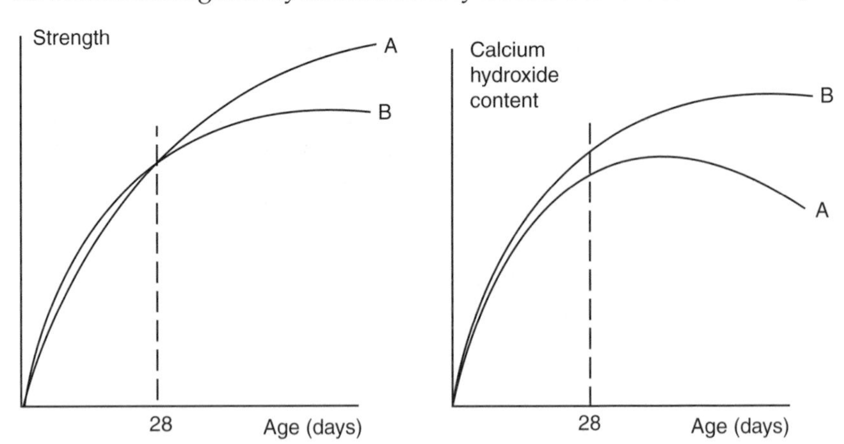

Fig. 3.10. Influence of calcium hydroxide on strength development (Cabrera and Plowman[35])

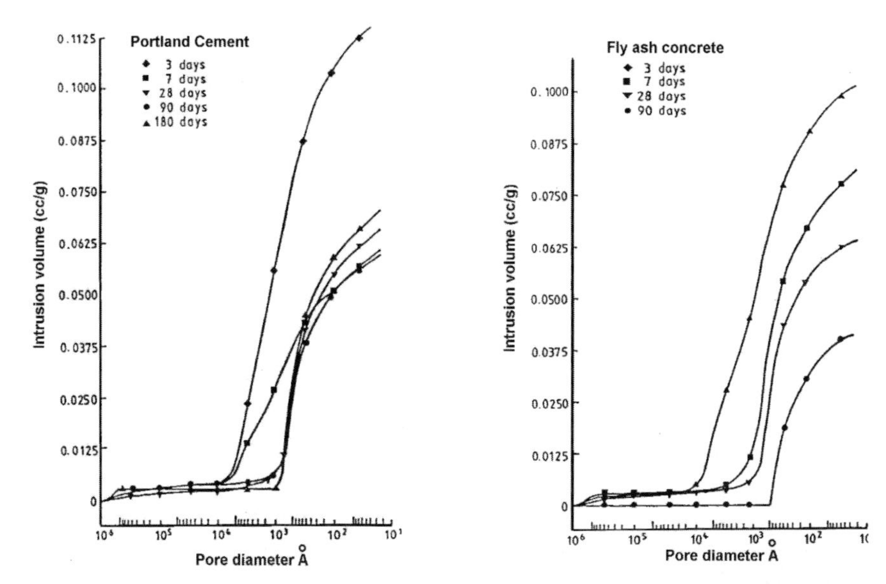

Fig. 3.11. Pore size distribution of ordinary Portland cement concrete (left) and fly ash concrete (right) at different ages (Ramezanianpour A, 1986, PhD thesis, University of Leeds)

a given cementitious content with increasing fly ash content lower early strengths are achieved. Taylor[39] explains the hydration processes involved in some detail. The presence of fly ash retards the reaction of alite, one of the components of cement, in the early stages of the reaction. However, alite production is accelerated in the middle stages owing to the provision of nucleation sites on the surface of the fly ash particles. The calcium hydroxide etches the surface of the glassy particles reacting with the SiO_2 or the Al_2O_3–SiO_2 framework. The hydration products formed reflect the composition of the fly ash with a low Ca/Si ratio.

The pozzolanic reaction gives a fly ash concrete its fine pore structure, low permeability, long-term strength gain properties and enhanced durability, since most durability problems are associated with the ingress of aggressive agents via the pore structure. Clearly, the finer the fly ash the greater the area exposed for reaction. In addition, the higher the temperature the greater the reaction rate. At later ages the contribution of fly ash to strength gain increases greatly, provided that there is adequate moisture to continue the reaction process.

Particle shape and density

Fly ash particles less than 50 μm are generally spherical, with larger sizes tending to be more irregular in shape. The spherical particles confer significant benefits to the fluidity of the concrete in a plastic state owing to the particle shape and by optimising the packing of particles. The fly ash spheres appear to act as 'ball bearings' within the concrete, reducing the amount of water required for a given workability. In general, the finer a fly ash the greater the water-reducing effect (Fig. 3.12[40]). A very coarse fly ash may not give any reduction in water demand, although this is dependent on the particle size distribution of the fine aggregates and cements in use. Visually, fly ash concrete may appear to be very cohesive until some form of compactive effort is applied, e.g. when compacted into the slump cone or vibrated. The reduction in water content reduces the propensity for bleeding and lowers the drying shrinkage potential.

When relatively coarse fly ash, i.e. 45 μm residue > 12%, is interground with clinker or ground separately, the water requirement of concrete is markedly reduced.[41] The grinding action appears to break down agglomerates and porous particles, but has little influence on fine glassy spherical particles, smaller than approximately 20 μm.[42]

The particle density of fly ash is typically 2300 kg/m^3, which is substantially lower than for Portland cement at 3120 kg/m^3. Therefore, for a given mass of Portland cement a direct, one-for-one by mass substitution of fly ash will give a greater volume of cementitious material. The mix design for the concrete should be adjusted in comparison to a Portland cement of the same binder content to allow for the increased volume of fine material by reducing the fine aggregate content. With very fine pozzolanic materials, a deflocculating agent, such as ordinary water-reducing plasticiser, can help to reduce the tendency of such fine fillers to agglomerate. This improves their relative water-reducing properties and aids the pozzolanic reactions by improving dispersal.

For a given 28 day strength the higher cementitious content needed in comparison with Portland cement concrete and lower water content required for fly ash-based concretes can give significantly higher quality surface finishes.

Variability of fly ash

Controlled fineness fly ash has been available in the UK since 1975 and over recent years the majority of fly ash/PFA supplied for concrete complies with BS 3829 Part 1, which limits the percentage retained on the 45 μm sieve to 12%. This standard by default restricts the variability of the PFA. In practice, in order to achieve the required fineness, PFA to BS 3892 Part 1 is classified using air-swept cyclones. The typical range of fineness found is ± 3%. BS EN 450 permits a wider range of variation of ± 10% on the declared mean fineness range of up to 40% retained on the 45 μm sieve. The typical range of fineness found is ± 5·0%.

The variability of the fineness is felt by many to influence adversely the variability of the concrete and reduce the commercial viability of using fly ash. However, Dhir *et al.*[43] concluded that LOI and fineness are only 'useful indicators' of fly ash performance. They show that the differing characteristics of seven cements produce a greater influence on concrete performance than the characteristics of eight differing fly ashes. Dhir *et al.* suggest that the fineness limits set in BS 3892 Part 1 are too low and that, a limit of 20–25% is more realistic. Matthews and Gutt[44] report on the effects of fineness, water reduction, etc., on a range of ashes with up to 18% retained on the 45 μm sieve. At a fixed W/C ratio, the range of strengths found indicates coefficients of variation of 17% at 365 days and 9% at 28 days. Typically, the coefficient of variation associated with a Portland cement is 5% at 28 days. The influence of cement characteristics on performance can be markedly reduced by the production of factory-produced cement, where the manufacturer can adjust fineness and ash level (within certain limits) to achieve the required concrete performance.

Fly ash variability must affect the final product somewhat and some restriction on the variability of fineness is needed. However, within British Standards many materials have minimal limits on variation, e.g. CEM I 42.5N can range between 42.5 and 62.5 MPa on a cement mortar prism. No quality-assured concrete manufacturer would take such highly variable cement, as it would have serious commercial implications. The same principle is true of fly ash. Commercial pressure would apply, e.g. for a highly variable fly ash only a low price could be commanded or expected.

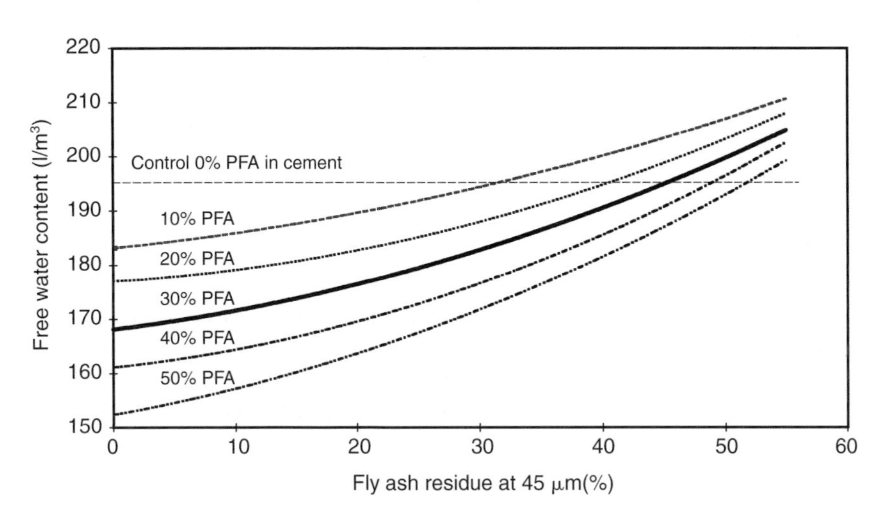

Fig. 3.12. Finer fly ash and/or more fly ash reduces water content (Owens[40]). The figure shows the effect of fly ash sieve residue on the water content of 50–75 mm slump concrete: nominal cement 300 kg/m³ with various percentages of fly ash

Fineness and pozzolanic activity

In general, fineness is an important factor because a smaller particle size means that a greater surface area will be exposed to the alkaline environment within the concrete. Fineness is normally assessed by the percentage retained on the 45 μm sieve (wet sieved) and typical values would range from a few per cent to 35%. With a single source of fly ash, the strength performance can be related to water demand and fineness (Fig. 3.12).[40] However, measuring the 45 μm residue of fly ash does not give much information about the very fine material/particle size distribution or structure. Cabrera and Gray[36] reported on the effects of particle size distribution of fly ash and the pozzolanic activity of such ashes. Various methods of comparing the surface area of the ashes were devised based on air permeability, nitrogen absorption (both dry and burnt), glycerol retention and iodine absorption. Cabrera and Gray conclude that there is no correlation between pozzolanic activity (determined from lime-ash mortars) and the calculation from particle size distribution (Table 3.6), except when using nitrogen absorption techniques. They proposed a system of classification based on the sphericity of the particles. When fly ash consists of spherical particles its measured surface area is very nearly equivalent to the surface area calculated from its particle size distribution. The two values increasingly differ with a higher proportion of irregular particles. Clearly, there are limitations to using the percentage retained on a 45 μm sieve as an indicator of the reactivity of fly ash, although it can be used as a measure of consistency.

BS EN 196-5 details a method of determining the pozzolanicity of a fly ash by comparing the hydroxide concentration of a mortar containing fly ash against the theoretical hydroxide content saturation possible. As part of a testing programme[29] carried out in 1999 the pozzolanicity for a number of UK sources was determined over a period of 20 weeks for both classified and unclassified fly ashes. The results of this work clearly show the

Table 3.6. Correlation coefficients between the strength of lime–sand–fly ash mortars and the fly ash specific surfaces as determined by different methods (reproduced from Cabrera and Gray[36])

Method of specific surface determination	Correlation coefficient		
	7 days	28 days	56 days
N_{dry}	−0.527	−0.673	−0.657
N_{burnt}	−0.055	−0.039	−0.058
Air permeability	−0.082	−0.090	−0.415
By calculation from particle size distribution	−0.043	−0.049	+0.349

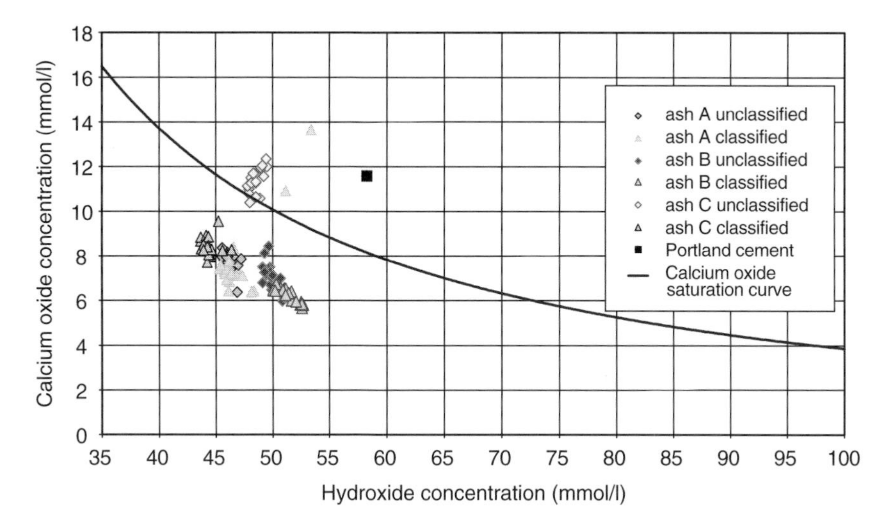

Fig. 3.13. Hydroxide concentration reduction due to pozzolanicity of various fly ashes: calcium oxide concentrations produced by three ashes, both classified and unclassified when tested with one cement, relative to saturation level at that pH

differences in reactivity from source to source (Fig. 3.13), as well as the improved reactivity of classified fly ash.

Cabrera and Hopkins looked at the fineness and reactivity of fly ash in the Drax project.[45] They carried out a detailed review of which factors influence the strength, porosity and workability of concretes. Nineteen sources of fly ash were used from power stations throughout the UK. A wide range of chemical and physical tests was carried out on the 720 samples of fly ash used in the project. The conclusions from this work can be summarised as follows:

- All qualities of fly ash could produce concrete of the desired strength if properly designed.
- The percentage retained on a 45 μm sieve is only partially applicable to the properties of the fly ash in concrete. Because of BS 3892 Part 1 requirements, over 70% of the fly ash available in the UK is deemed unsatisfactory without classification. However, in reality coarser fly ash does not have the same status or economic benefit and yet its use would be beneficial to the durability of the concrete in the longer term.
- The important particle sizes are 10·1, 4·8 and 3·0 μm. These govern fly ash performance. A shape factor system was proposed to classify fly ash more accurately. LOI was also found to play an important role.
- The use of a plasticising admixture (a lignosulfonate-based water-reducing admixture) reduced the effect on the strength of the variability in the fly ash.

Bogdanovic[46] compared Portland cement concrete against fly ash concretes with between 58·9% and 63·1% retained on the 45 μm sieve. The mixes had roughly equal 28 day strength. At 112 days, although the fly ashes used were very coarse, the fly ash concretes achieved strengths some 17% higher than the Portland cement-only concrete. The modulus of elasticity was reduced by 8% at 112 days. Allen[47] reported on Lednock dam, the first fly ash structure of any size built in the UK. Although problems with the variability of LOI and fineness were reported, the dam has proven to be durable. Bouzoubaa *et al.*[47] reviewed the properties of fly ash cements produced by intergrinding with clinker and by separate grinding. It was concluded that the grinding of fly ashes increased their specific gravity and fineness, and consequently reduced the water requirement and increased their pozzolanic activity.

It is clear that characterising the reactivity of fly ash is somewhat complex. The pozzolanicity depends on the surface area of fly ash exposed to the pore solution, the alkalinity of that pore solution, the curing temperature and the chemical composition of the fly ash, e.g. the source and type.

Fineness and workability/strength characteristics

The particle shape and finer fractions of fly ash are capable of reducing the water content needed for a given workability (Fig. 3.12). These effects are felt to be due to void filling on a microscopic scale replacing water within the concrete mix. Dewar[48] found results from his mix design system which correlate with water reductions found in practice when using fly ash. Within the Dewar system, the mean particle size is the governing factor that cannot be properly expressed using a single point test such as retention on the 45 μm sieve. As the surface area of a spherical particle is proportional to the radius squared, small quantities of the finer fractions can have a highly significant effect on the water demand. Where a water reduction is found this partially contributes to the relative strength performance of the cement/fly ash combination by acting as a solid particulate plasticiser.

The strength performance of a fly ash concrete will depend on the water reduction achieved and the pozzolanic performance of the cement/fly ash combination. While finer fly ashes are presumed to affect strength significantly, not all researchers find this true. Brown[49] concluded that 'for curing at 50°C a fine ash makes a larger contribution to strength than does a coarse grade but the difference is only about 2 MPa for each 10% fly ash/(fly ash + cement) by weight', and '... that the selection of either fly ash or cement has no greater significance with regard to strength'. The fly ashes were wide ranging, with 12% and 32% retained on the 45 μm sieve. Such effects may be simply physical, due to the particle-packing properties of the concrete mix constituents, e.g. the fine and coarse aggregates.

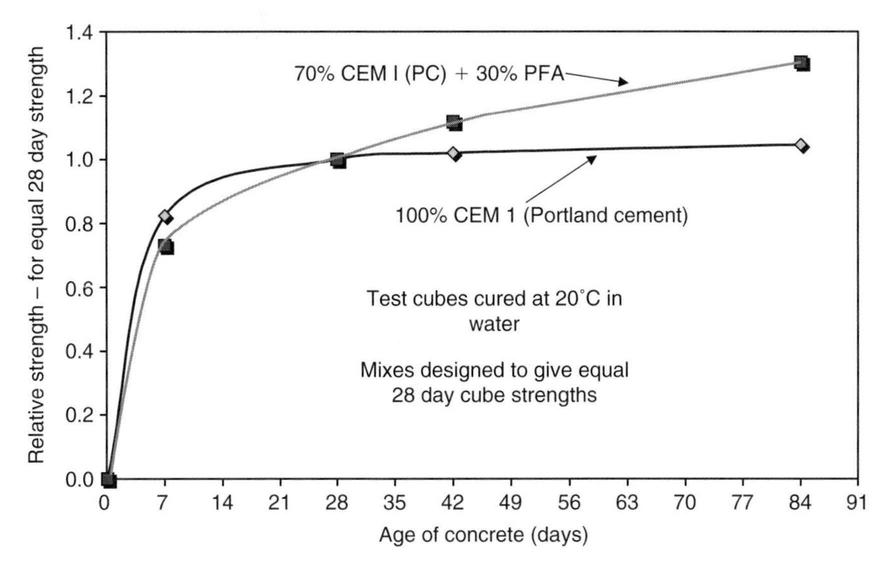

Fig. 3.14. At 20°C concrete containing 30% fly ash continues to gain significant strength

As with the pozzolanic activity, there is a trend towards a reduction in water requirement with increasing fly ash fineness. However, the strength performance may not be directly related to the fineness. Fly ash chemistry and surface area are the controlling factors. Fly ash affects the rate of gain of strength in concrete. At early ages, the rate of gain of strength is lower than an equivalent Portland cement concrete of similar grade. In the long term, however, it may be higher. Mix design and in particular W/C ratios play an important part in strength development. Typical strength development graphs of standard cubes with and without fly ash are shown in Fig. 3.14. The figure also clearly illustrates the effect of inclusion of fly ash in concrete on the development of strength at early ages. Consideration should always be given to this effect.

Where higher curing temperatures are encountered, as in thick sections, significantly higher *in situ* strength can be achieved than in test cubes cured at 20°C. Figure 3.15 shows this effect for some 30 MPa grade, 1·5 m cubic concrete specimens.[50] These were insulated on five sides to re-create thick concrete sections. The elevated temperatures enhanced the pozzolanic reactions and the *in situ* strengths for the concrete containing fly ash.

Heat of hydration

The development of concrete mix design has seen an increase in the proportion of cement being replaced by fly ash. Early uses of fly ash in

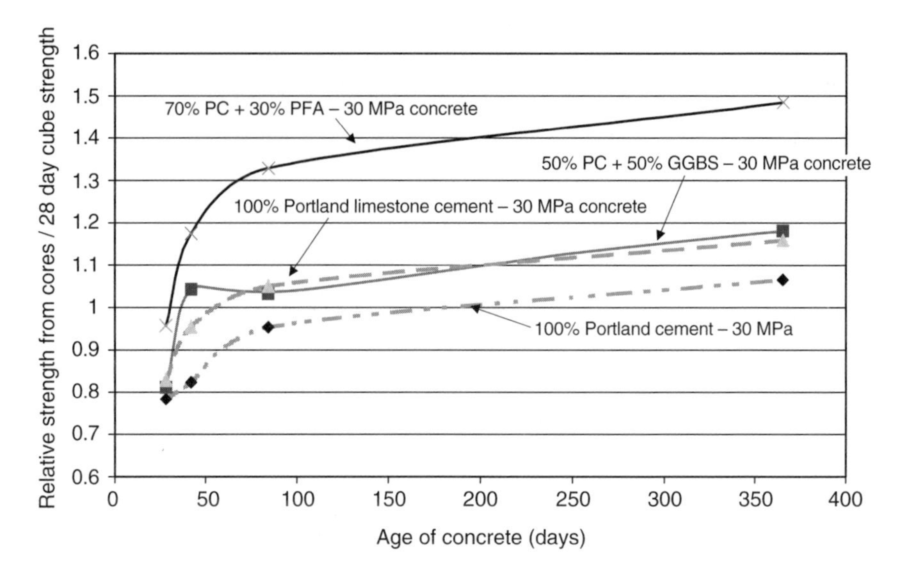

Fig. 3.15. The long-term in situ *strength of concrete may be improved significantly by 30% pulverised fuel ash (PFA) relative* in situ *strengths of various cement types (gravel concrete mixes, cast in summer, 1·5 × 1·5 × 1·5 m blocks)*

concrete concerned reducing the heat evolution in hardening concrete. In particular, for mass concrete in large dams it was found that by replacing a proportion of the cement with fly ash a large reduction in heat was achieved. A measure of this reduction in heat can be seen for similar strength concretes in Fig. 3.16. This lowered the potential for thermal cracking and produced a less porous structure.

The hydration of cement compounds is exothermic, with up to 500 J/g being liberated. Concrete is a poor conductor of heat, with the result that the temperature at the interior of a concrete mass will rise significantly during the hydration cycle. At the same time, external surfaces will be cooled by ambient temperatures and damaging temperature gradients may occur, resulting in cracking in the section.

The introduction of fly ash to replace a proportion of cement in concrete influences the temperature rise during the hydration period. The rate of the pozzolanic reaction increases with increasing temperature; however, the peak temperatures in fly ash concrete are significantly lower than in equivalent Portland cement concretes. Figure 3.17 shows adiabatic temperature curves (°C/100 kg of cement) for a range of total cement contents.

Setting time and formwork striking times

The inclusion of fly ash in concrete will increase the setting time compared with an equivalent grade of Portland cement concrete. There is

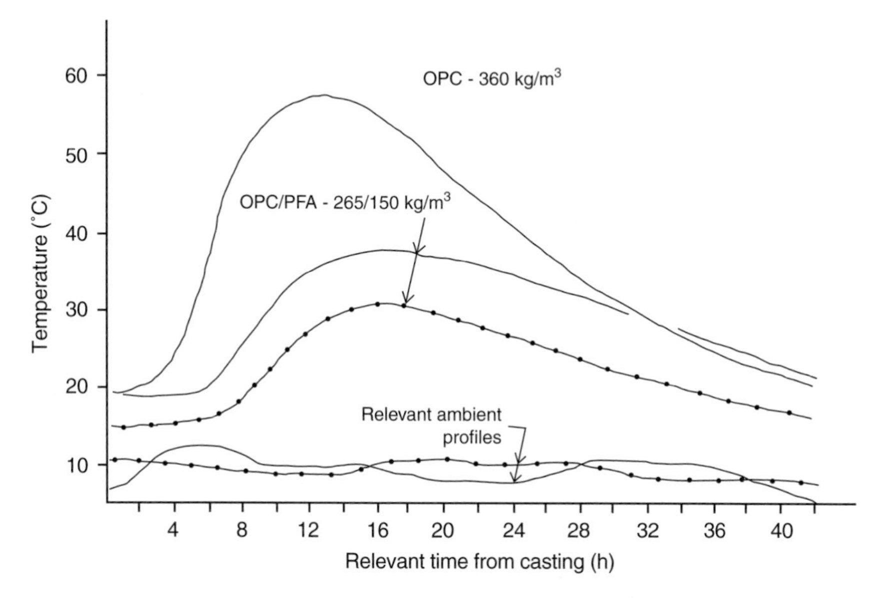

Fig. 3.16. Heat of hydration with time (Woolley and Conlin[51])

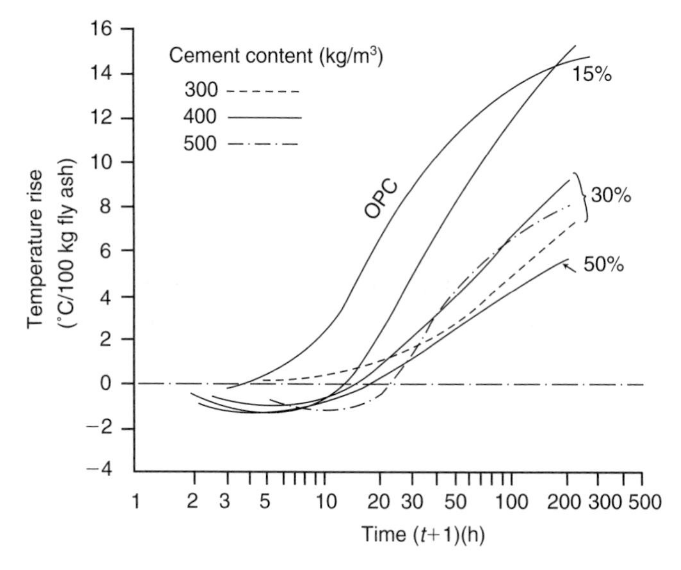

Fig. 3.17. Temperature rise in fly ash and Portland cement (PC) concretes (Bamforth[52])

undoubtedly a delay in the onset of the hydration of fly ash concrete, but it has been shown by Woolley and Cabrera[53] that the actual gain in strength once hydration has started is greater for fly ash concrete at normal temperature regimes. When 30% fly ash is used to replace Portland cement in a mix, the setting time may be increased by up to 2 h. This increased setting time reduces the rate of workability loss. However, it may result in practical difficulties for finishing, particularly during periods of low temperature. In compensation, it will reduce the incidence of cold joints in the plastic concrete. EN 197-1 imposes initial and final setting times that are comparable with CEM I requirements.

Formwork striking times at lower ambient temperatures normally will need to be extended in comparison to Portland cement concrete, especially with thin sections. In practice, vertical formwork striking times can be extended without this affecting site routines, e.g. the formwork is struck on the following day. For soffit formwork, greater care has to be taken. Reference should be made to BS 8110[54] for recommended striking times. Temperature-matched curing can be used to ensure that sufficient *in situ* strength has been achieved while allowing for the concrete curing conditions.

Elastic modulus

The elastic modulus of fly ash concrete is generally equal to or slightly better than that for an equivalent grade of concrete. The direct relationship between elastic modulus and strength is seen in Fig. 3.18 for concrete cured at different regimes. The figure shows the slower rate of gain in

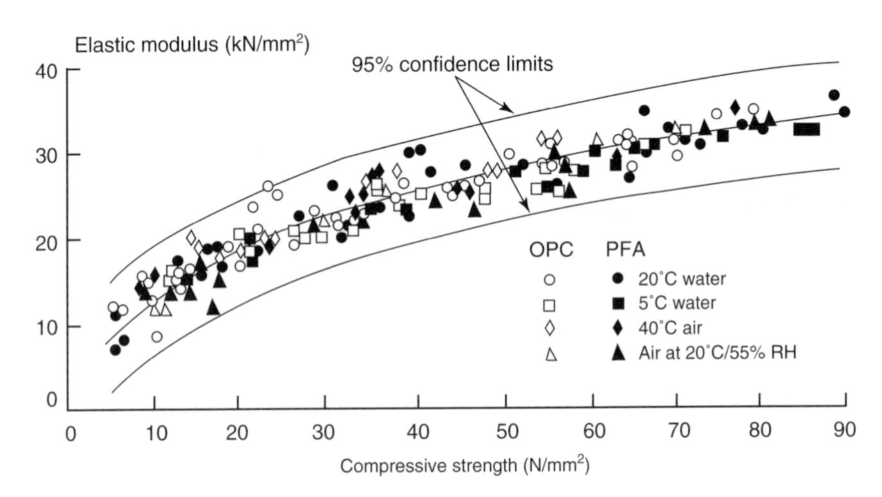

Fig. 3.18. Relationship between the modulus of elasticity and strength for fly ash (PFA) and Portland cement (OPC) concretes (Dhir et al.[55]*)*

strength experienced with fly ash concrete followed by the ongoing development in strength. In a similar manner at early ages, the elastic modulus of fly ash concrete is marginally less than for equivalent Portland cement concretes at later ages.

Creep

The greater strength gains of fly ash concretes have shown lower creep values, particularly under conditions of no moisture loss. These conditions may be found in concrete remote from the cover zone of a structure. Where significant drying is permitted the strength gain may be negligible and creep of ordinary Portland cement and fly ash concretes would be similar. Figure 3.19 shows the creep of fly ash and Portland cement concrete loaded to different stress levels.[55]

Tensile strain capacity

The tensile strain capacity of fly ash concretes has been found to be marginally lower than for Portland cement, and fly ash concretes exhibit slightly more brittle characteristics.[56] There is possibly a greater risk of

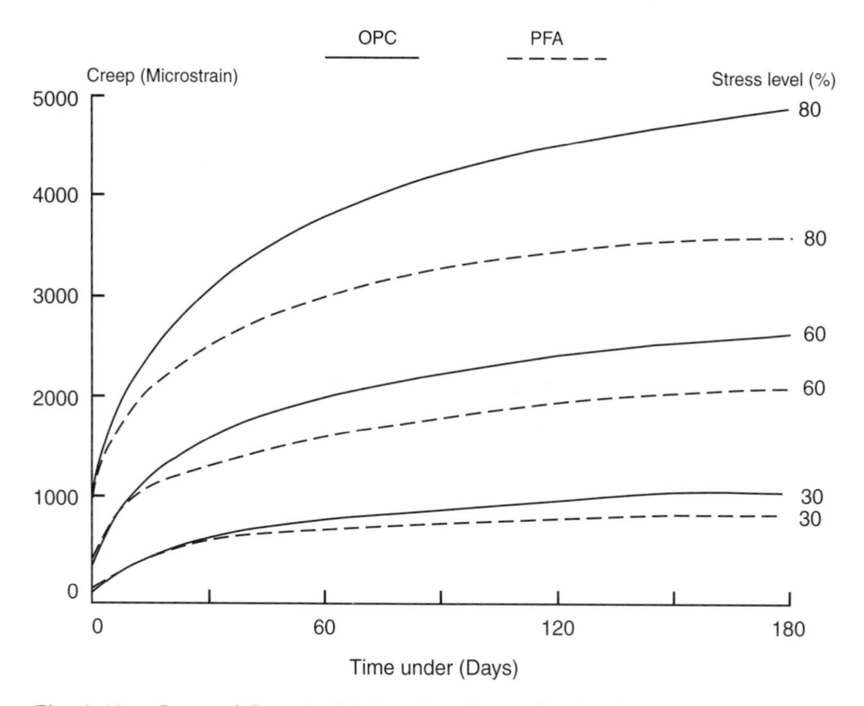

Fig. 3.19. Creep of fly ash (PFA) and ordinary Portland cement (OPC) concrete loaded to different stress levels (Dhir et al.[55]*)*

early thermal cracking for given temperature drop, partially offsetting the benefits of lower heat of hydration in the fly ash concrete.

Coefficient of thermal expansion

The coefficient of thermal expansion of concrete is largely determined by the type of coarse aggregate used. Replacement of a proportion of cement with fly ash will have little influence on this property.[57]

Curing

The hydration reaction between cement and water provides the mechanism for the hardening of concrete. The degree of hydration dictates strength development and all aspects of durability. If concrete is allowed to dry out hydration will cease prematurely. Fly ash concrete has slower hydration rates and the lack of adequate curing will, as with other concretes, affect the final product. Thin concrete sections are more vulnerable than constructions of thicker section, where heat of hydration will promote earlier hydration.

Durability of concretes made with fly ash

Keck and Riggs[58] quote Ed Abdun-Nur, who said, 'concrete, which does not contain fly ash, belongs in a museum'. They list the benefits of using fly ash, e.g. low permeability, resistance to sulfate attack, alkali silica reaction and heat of hydration, giving references to their conclusions. Thomas[59] reports that cores taken from 30-year-old structures in general have performed well; such structures must have been made with unclassified fly ashes over which minimal control of fineness, LOI, etc., would have been exercised. Berry and Malhotra[60] review the performance of fly ash concretes in detail. They show the beneficial properties of using fly ash for sulfate resistance, chloride penetration, permeability, etc. Some specific durability aspects are considered next.

Penetration of concrete by fluids or gases may adversely affect the durability. The degree of penetration depends on the permeability of the concrete, and since permeability is a flow property it relates to the ease with which a fluid or gas passes through it under the action of a pressure differential. Porosity is a volume property, representing the content of pores irrespective of whether they are interconnected, and may or may not allow the passage of a fluid or gas.

There are two mechanisms by which liquid or gas ions are transported in concrete:

- Adsorption is the process whereby molecules collect in a condensed form on the surface of concrete. Absorption describes the way in which concrete takes in a liquid to fill voids or pores within the material.

- Diffusion describes the process by which liquid or gas ions pass through concrete under the action of a concentration gradient. The rate at which liquid or gas ions pass through concrete owing to a concentration gradient is known as diffusivity.

Other important factors are

- the moisture condition of concrete has a major influence on values of absorption, flow and/or diffusion[61];
- during the hydration process gel products are precipitated and block pores in the concrete, reducing flow and/or diffusion.

Alkali silica reaction

The alkali silica reaction (ASR)[62] is potentially a very disruptive reaction within concrete. However, the amount of damage that has occurred is small in comparison with the amounts of concrete produced. The first reported occurrence in the UK was in 1976 and by 1983 some 50 cases were known. ASR involves the higher pH alkalis such as sodium and potassium hydroxides reacting with certain forms of silica, usually within the aggregates, producing gel. This gel has a high capacity for absorbing water from the pore solution, causing expansion and disruption of the concrete. Some greywacke aggregates are particularly susceptible to ASR. The main source of the alkalis is usually the Portland cement or external sources such as cleaning fluids containing sodium hydroxide. Fly ash contains some sodium and potassium alkalis but these are mainly held in the glassy structure and therefore are not available for reaction. Typically, only some 16–20% of the total sodium and potassium alkalis in fly ash are water soluble.

Many researchers have found that fly ash is capable of preventing ASR. The glass in fly ash is in a highly reactive fine form of silica. It has been found that the ratio of reactive alkalis to surface area of reactive silica is important in ASR. A pessimum ratio exists where the greatest expansion will occur. However, by adding more reactive silica the dilution of the reaction with the alkalis, coupled with the low permeability of fly ash concrete, effectively means that no disruptive reaction happens. The recommendations[63] within the UK require a minimum of 25% BS 3892 Part 1 fly ash to prevent ASR. For coarser fly ashes, a minimum of 30% fly ash may be required to ensure sufficient surface area to prevent ASR. Small quantities of fly ash with low reactivity aggregates and a source of alkalis may be more susceptible to ASR if the pessimum silica–alkali ratio is achieved. Even when total alkalis within the concrete are as high as 5 kg/m^3 fly ash has been found[64] to be able to prevent ASR. The addition of fly ash reduces the pH of the pore solution to below 13, at which point ASR cannot occur. The use of low alkali cements has a similar effect. However, the detailed mechanisms by which fly ash prevents ASR are complex and imperfectly understood.

According to Taylor[65] the mechanism by which fly ash reduces the risk of ASR can be summarised as

- the fly ash pozzolanic reaction is similar to the alkali silica reaction
- pessimum proportions of SiO_2/Na_2O must exist for disruptive ASR to occur
- as the Ca/Si ratio decreases the alkali cations are more readily taken up by the SiO_2
- more C–S–H hydration products are formed rather than expansive gel.

The *ACI manual of concrete practice*[66] suggests few restrictions on the effectiveness of fly ashes. It states that 'The use of adequate amounts of some fly ashes can reduce the amount of aggregate reaction'. It is later suggested that ashes only have to comply with ASTM C618, which permits a wide range of fly ashes. Fournier and Malhotra[67] investigated the ability of a range of fly ashes to prevent ASR. The AI was found to affect the ASR performance; however, there was no correlation between fineness and AI. Nant-y-Moch, Dinas and Cwm Rheidol dams are excellent examples of fly ash preventing ASR. The Nant-y-Moch and Cwm Rheidol dams were constructed using 'run of station' fly ash from Bold power station and the structures have proven to be durable. These dams have performed well, in comparison with the Portland cement-only Dinas dam, which has some evidence of ASR cracking and yet was built around the same period using the same aggregates.

Carbonation

The ingress of CO_2 into concrete and the subsequent conversion of lime to carbonate reduce the pH of the matrix to about 9. To occur this mechanism requires two factors, some moisture in the concrete, but not saturated, and a path by which the CO_2 can diffuse through concrete. This reaction is not detrimental to the concrete as such; in fact, it may help to reduce permeability and improve sulfate resistance, but is deleterious to the reinforcing steel in reinforced concrete. The high pH found in normal concrete maintains a passivity layer on the reinforcements which prevents corrosion. As fly ash pozzolanically reacts with lime, this potentially reduces the lime available to maintain the pH within the pore solution. However, fly ash reduces the permeability of the concrete dramatically when the concrete is properly designed and cured. In addition, when designing concretes for equal 28 days strength the slow reaction rate of fly ash usually means that the total cementitious material is increased. This increase partially compensates for the reduction in available lime. Coupling this with the lower permeability leads to the result that the carbonation of fly ash concrete is not significantly different from Portland cement-only concrete of the same grade (28 days).[68] The *ACI manual of concrete practice*[66] confirms this view: 'Despite the concerns that the pozzolanic action of fly ash reduces the pH of concrete researchers have found that an alkaline environment very similar to that in concrete without fly ash remains to preserve the passivity of the steel'.

High Marnham power station was built in 1957. A 275 kV substation was built with some of the bases using concrete where a partial replacement of the Portland cement with fly ash was made. As the original concrete was well documented this was a rare occasion on which the long-term performances of Portland cement-only and fly ash concretes could be assessed. In 1985, Cabrera and Woolley[69] reported their findings after a detailed study. The fly ash concrete contained 20% of the cement content as 'run of station' fly ash, with an estimated 16·5% retained on the 45 μm sieve. They reported that this concrete had no significant depth of carbonation. The durability of the concrete was excellent, with the compressive strength development of the fly ash concrete being twice that of the Portland cement-only concrete.

It is clear that carbonation is a complex function of permeability and available lime. With properly designed, cured and compacted fly ash concrete, carbonation is not significantly different from other types of concrete. With extended curing and the low heat of hydration properties of fly ash concrete, the resulting low permeability may more than compensate for the reduced lime content.

Sea water and chloride ingress

Chlorides penetrating concrete will attack the reinforcing steel and cause corrosion. An electrochemical cell is formed within the concrete between the reinforcing steel and the pore solution. The Fe^{3+} ions from the passivity layer of Fe_2O_3 pass into solution, while the electrons pass along the reinforcing bar. They recombine to form ferric hydroxide [$Fe(OH)_3$]. The reaction requires oxygen, water and the presence of chlorides. The corrosion at the anode site leads to loss in section of the reinforcing steel and can ultimately lead to failure of the element.

Chlorides can come from many sources. First, it is important to restrict the chloride content in the constituents of the concrete to a minimum and guidance is given within various standards, e.g. BS 5328, BS 8110 and BS 8500. Chlorides from external sources are normally from seawater and deicing salts used on roads. Tricalcium aluminate (C_3A), a compound found in Portland cement, is able to bind chloride ions, forming calcium chloroaluminate. Similarly, tetracalcium alumino ferrite (C_4AF) can also reduce the mobility of chloride ions, forming calcium chloroferrite. Fly ash also contains oxides of alumina, which are able to bind chloride ions.

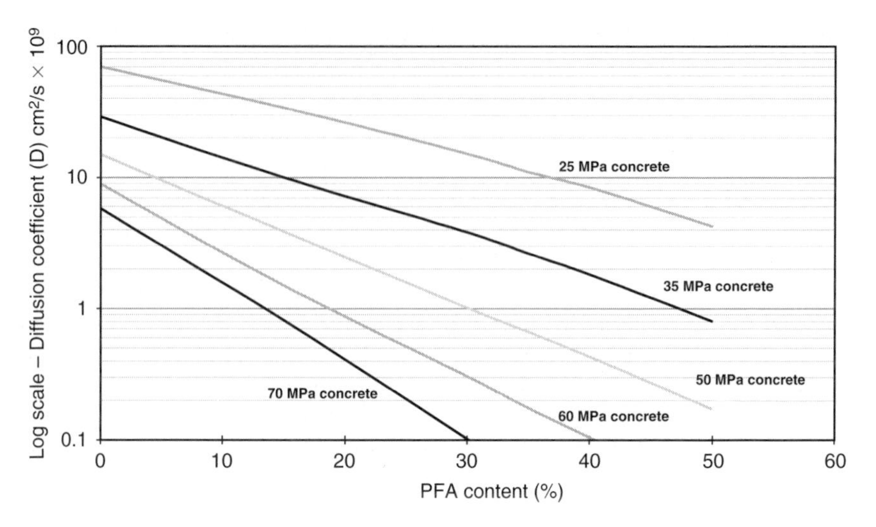

Fig. 3.20. Estimate of chloride diffusion coefficient: by increasing the strength and fly ash content of concrete the permeability reduces significantly (Owens[40])

One highly effective way to reduce chloride ingress is to lower the permeability of the concrete. When fly ash is added to concrete, the permeability is significantly lower (Fig. 3.20). This reduces the transportation rate of chloride ions, and many other materials, into the concrete.

Freeze–thaw properties

The freezing of water within concrete is responsible for gradual but severe damage to the concrete. The volume of water increases by around 9% when freezing occurs in the pore structure. If the ice is unable to escape, the pressure exerted on the concrete simply exceeds the tensile strength of the concrete. This initially manifests itself as microcracking, but after being subjected to many cycles of freezing and thawing, continuous cracking leads to failure. This is seen as spalling of the surface and ultimately there is loss of cover to the reinforcing steel, corrosion and, finally, failure of the concrete.

Fly ash concrete of the same strength has a similar resistance to freeze–thaw attack as Portland cement concrete. Dhir *et al.*[70] reported on the freeze–thaw properties of concrete containing fly ash. They used nine fly ashes of various sieve residues. All mixes were designed to give equal 28 day strengths. Freeze–thaw was assessed using an 8 h cycle of 4 h at −20°C and 4 h at +5°C. Ultrasonic pulse velocity (UPV) and changes in length were used for the assessment of freeze–thaw performance. They found, as have other researchers, that adding fly ash reduces freeze–thaw resistance unless air entrainment is used. However, Dhir *et al.* found that even 1·0% of air entrained in the concrete gave superior performance compared with plain concrete, irrespective of the cement type.

It is clear that the less permeable and denser the mortar matrix is within the concrete the less space is available to relieve the pressures associated with the expansion of freezing water. The additional hydration products created by the pozzolanic reaction result in fly ash being less frost resistant unless voids are artificially created by the addition of air entrainment. In all low-permeability concretes, irrespective of the fineness or chemistry of the fly ash or other constituents, freeze–thaw resistance is given by air entrainment. The air bubble structure is able to relieve expansive stresses of the freezing water, presuming proper curing regimes have been adhered to. The importance of good concrete practice, the correct mix for the application, adequate compaction, curing, etc., cannot be over emphasised.

Sulfate resistance

Sulfates in solution attack the hardened cement in concrete. The attack is both chemical and physical. Sulfate ions react with hydrated calcium aluminates to form ettringnite and they combine with free calcium hydroxide to form gypsum. The rate of attack is influenced by concentration and type of sulfate, i.e. calcium sulfate and magnesium sulfate. The pH of soil or groundwater, water table and mobility of groundwater, and the concrete constituents, compaction and permeability are all influences in this process. Considerable increases in volume and disruption of the hardened concrete result from the expansive reactions to form ettringnite and gypsum.

Fly ash concrete can increase the resistance to sulfate attack compared with a CEM I concrete of similar grade. Deterioration due to sulfate penetration results from the expansive pressures originated by the formation of secondary gypsum and ettringnite. The beneficial effects of fly ash have been attributed to a reduction in pore size slowing the penetration of sulfate ions. Less calcium hydroxide is also available for the formation of gypsum.

The smaller pore size of fly ash concrete reduces the volume of ettringnite that may be formed. One of the major constituents of cement that is prone to sulfate attack, tricalcium aluminate (C_3A), is diluted since a proportion of it will have reacted with the sulfates within the fly ash at an early age. Building Research Establishment[71] Digest 363 discusses the factors responsible for sulfate and acid attack on concrete below ground and recommends the type of cement and quality of curing to provide resistance. Concrete made with combinations of Portland cement and BS 3892 Part 1 fly ash, where the fly ash content lies between 25% and 40%, has good sulfate-resisting properties and may be used for up to class 4A (Digest 363 classification) sites. Fly ash concretes do not appear to perform well in the presence of magnesium sulfates, e.g. class 4B. When Portland-pulverised fuel ash cements to BS 6588 are used, the cements are regarded as sulfate resisting only when the fly ash content is not less than 25%.

Sea water contains sulfates and attacks concrete through chemical action. Crystallisation of salts in pores of the concrete may also result in disruption. This is a particular problem between tide marks subject to alternate wetting and drying. The presence of a large quantity of chlorides in seawater inhibits the expansion experienced where groundwater sulfates have constituted the attack.

Laboratory studies by the Building Research Establishment over a 5-year period led to the 1991 revision to BRE Digest 363, 'Sulfate and acid resistance of concrete in the ground'. Fly ash concrete samples were immersed in various sodium sulfate and magnesium sulfate solutions. Four different fly ashes and three different cements were used and constituent proportions were varied. The study found that concrete containing combinations of fly ash with Portland cement, even when the C_3A content was as high as 14%, can be compared with concrete containing sulfate-resisting cement immersed in sulfate solutions equivalent to class 4 exposure. Blends of fly ash with sulfate-resisting Portland cement (SRPC) generally gave better sulfate resistance than SRPC alone at a cement content of 400 kg/m^3. At a lower cement content, when concrete was exposed to 1·5% sodium and magnesium solutions, SRPC gave better results.

Thaumasite

Much has been said about thaumasite in the UK, resulting from the discovery of a number of damaged structures on a motorway. Bensted[72] describes the chemistry of the thaumasite reaction. Thaumasite attack occurs at temperatures below 15°C and requires the presence of calcium carbonate. Limestone aggregates are a source of calcium carbonate, with oolitic limestone appearing to be the most reactive form. Sulfates react with the calcium carbonate and the C_3S and C_2S hydrates, forming thaumasite. As these are the strength-giving phases of the cement, their removal results in the disintegration of the concrete to a white, powdery, sludge-like material. This reaction is not expansive and may not be easily detected below the ground. At the time of writing several research projects remain outstanding which should answer some of the questions posed by this reaction such as:

- The affected motorway structures in the UK were backfilled with pyritic clay. When exposed to air the pyrites decompose to sulfuric acid. The significance of acid attack has yet to be determined.

- At 5°C thaumasite forms readily in the laboratory. However, the incidence of thaumasite attack, other than when pyritic clay and coal shales have been used as a back fill, is relatively small. Why?
- Cement type has a bearing on the degree of attack. GGBS appears to prevent the reaction at higher cement contents and low W/C ratios with limestone aggregates. However, the mechanisms are not fully understood as thaumasite reactions have been detected when siliceous aggregates and GGBS are used.

In 1980, Burton[73] carried out a series of mixes to determine the relative performance of concretes made with Portland cement, SRPC and Portland/fly ash-blended cement concretes when exposed to sulfate attack. The sulfate solutions were not heated and the tanks were in an unheated, external storeroom. The Portland cement mixes all showed signs of deterioration after 6 months in magnesium sulfate solution but only slight deterioration in sodium sulfate solution. However, after 5 years' immersion there was total disintegration of all Portland cement samples. The oolitic limestone mixes suffered between 22·2% and 42·5% weight loss after 1 year in magnesium sulfate and the crushed carboniferous limestone mixes 14·3–27·3% weight loss (Fig. 3.21). This indicates the effect of aggregate type on the rate of deterioration. Although thaumasite was not well understood at the time of the project, a re-evaluation of the data and photographs suggests that the deterioration found with both the Portland cement and SRPC mixes was probably due to the thaumasite form of sulfate attack.

Bensted[72] believes that the addition of fly ash should reduce the problems with thaumasite, simply because the permeability of fly ash concrete should prevent sulfate in solution from penetrating to any depth.

Fig. 3.21. Deterioration of various cementitious types in sulfate solution (Burton[73]). OPC: Portland cement; SRPC: sulfate-resisting Portland cement

Resistance to acids

All cements containing lime are susceptible to attack by acids. In acidic solutions where the pH is less than 3·5, erosion of the cement matrix will occur. Moorland waters with low hardness, containing dissolved CO_2 and with pH values in the range 4–7, may be aggressive to concrete. The pure water of melting ice and condensation contain CO_2 and will dissolve calcium hydroxide in cement, causing erosion. In these situations, the quality of concrete assumes a greater importance.

Concrete mix design

Strength effects

Concrete mixes for early constructions were designed to replace 20% by mass of Portland cement with an equal mass of fly ash. There is a risk of creating problems if very small percentages of fly ash are used in concrete, e.g. there is an increased risk of ASR with susceptible aggregates. The principle of a mass-for-mass replacement of Portland cement depresses early and 28 day strength relative to ordinary Portland cement concretes, and takes no account of workability. To overcome this it was shown that by replacing some of the fine aggregate with fly ash and increasing the cementitious content, equal 28 day strengths could be achieved.[74] Smith[17] developed a method based on applying a cementing efficiency factor known as the *k*-factor. The mix design was adjusted as given below:

$$W/C_f = W/(C+kF), \tag{3.2}$$

where W/C in the equation for plain Portland cement concrete is replaced by the adjusted W/C_f ratio; W = weight of water, C = weight of Portland cement, C_f = equivalent weight of Portland cement, k = cement efficiency factor for fly ash, and F = weight of fly ash.

k-Factors can be created for and applied to many purposes, e.g. equal 28 day strength, equal chloride diffusion or equal durability. A calculated *k*-factor will change depending on the Portland cement source, the curing temperature and conditions, the fly ash source, etc. The technique has been corrupted in that a single *k*-factor for CEM I 42·5 N of 0·40 is being used to adjust minimum cement content in European standard EN 206. As Smith[17] used the W/C ratio, this has prejudiced the use of EN 450 fly ash and water-reducing admixtures. Within the UK, National Standards will continue to permit classified PFA to BS 3892 Part 1 to be fully counted towards the cement content. Similarly, EN 450 fly ashes will also be permitted to count fully towards the cement content if additional performance testing is carried out.

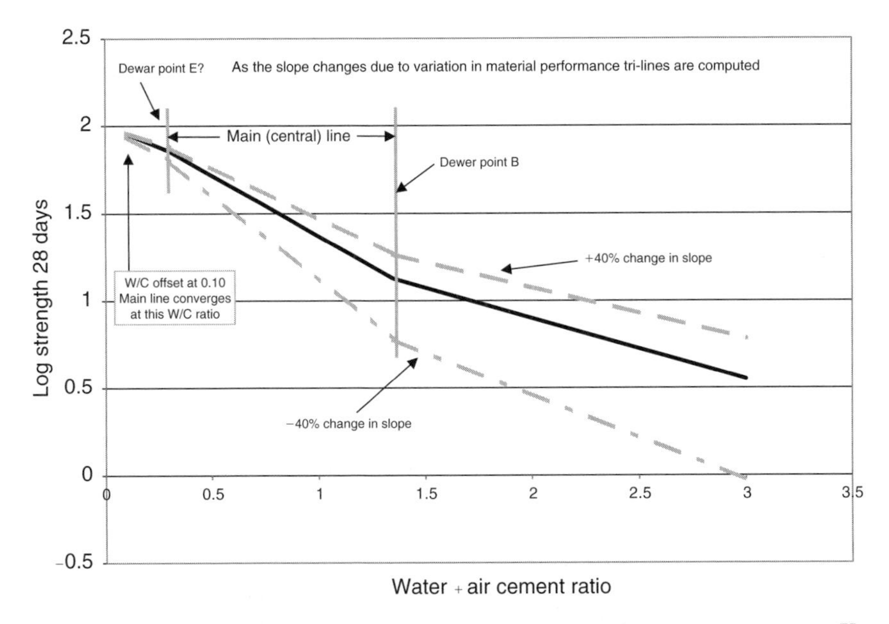

Fig. 3.22. Abram's law over a wide range of water/cement (W/C) ratios (Sear[75]). Families from the tri-line Abrams law approach; based on PC, 50 mm slump concrete with no admixtures

The k-value approach applies a single point value to a highly variable material such as concrete. To control the quality of concrete properly requires a system of continual monitoring. Duff Abrams established a law in 1919 which relates strength to W/C ratio as given below:

$$S = K_1/(K_2^{W/C}), \tag{3.3}$$

where S = strength, K_1 and K_2 = constants, W = mass of water, and C = mass of cement.

This rule can form the basis of a quality-control system as shown by Sear[75] that can control, compensate and even predict the strength performance of concrete mixes. A triple linear relationship can be shown to exist as in Fig. 3.22, which can relate a wide range of W/C ratios to strength. Typical strength versus cement content curves are shown in Fig. 3.23 for Portland cement and a fly ash concrete.

As the durability of concrete is often related to the strength or W/C ratio, or both, such systems offer better control over the final material than the broad-brushstroke approach of the k-value.

Mix proportioning

There are many methods of designing concrete mixes.[76] Most give an indication of the optimum proportions required and resort to laboratory trial

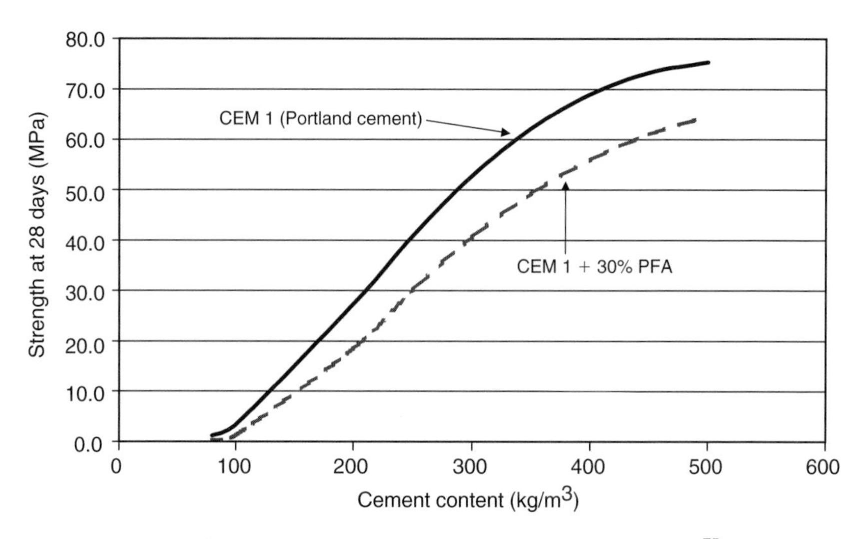

Fig. 3.23. Typical cement content versus strength relationship (Sear[75])

mixing as the final step in verification. Because of the particle shape and fineness of fly ash the design of concrete mixes is slightly different from normal Portland cement concretes in order to obtain the greatest technical and economic benefits from the mix. The spherical shape of the particles reduces the water content of the mix and the lower particle density in relation to Portland cement leads to a greater volume of cementitious fines in the mix. As most fine aggregates are washed to remove excessive fine material, fly ash can often correct for minor deficiencies in the overall grading of the fine aggregate/cementitious material, reducing the water content further.

Fly ash mixes tend to be somewhat more cohesive than Portland cement mixes and some compensation in the mix design is beneficial rather than a direct weight-by-weight replacement. In general, an increase in the coarse content of 3% based on the coarse/fine ratio will correct for this. Table 3.7 indicates two mixes of equal 28 day strength.

For most applications, 30% of the cementitious material is PFA/fly ash, which is the norm. However, for low heat applications, economic reasons and for especially demanding durability requirements, higher fly ash usage rates may be used. EN 197-1 permits CEM IV-B cements to contain up to 55% fly ash as a percentage of the nucleus, which is equivalent to 52·3% fly ash.

Fly ash reduces the rate of bleeding owing to the lower permeability, increased cohesiveness and lower water content of fly ash concrete. This is an aid when designing high workability or self-compacting concrete mixes to reduce settlement and cracking problems. However, efficient curing is very important. Excessive evaporation of water that is greater than the rate

Table 3.7. Adjusting a Portland cement (PC) to a 30% fly ash (PFA) mix for equal grade at 28 days

Typical material contents in kg/m^3 for a grade 40 MPa concrete at 28 days of fixed workability	PC-only version	PC plus 30% BS 3892 Part 1 PFA version	Comments
Total cement content	325	365	The total mass of the combination normally increases by approx. 12%
PC	325	255	This blend ratio is based on 30% PFA of the combination being used. For coarser fly ash a greater total cement content or smaller proportion of ash may be needed to maintain equal strength
PFA		110	
20/5 mm gravel	1200	1238	The coarse/fine ratio of the PC mix is 36·5% and 33·3% for the 30% PFA version. Increasing the coarse aggregate content of the PFA mix compensates for the cohesive nature of PFA
Sand	691	617	
Water content	164	154	With PFA, a water reduction is normally found. Generally, for equal workability, it is found that for every 10% of BS 3892 Part 1 fly ash added the water content may be reduced by 3%.[77] With increasing coarseness of the PFA, this benefit may not be seen
Mix density (kg/m^3)	2380	2374	Although there is less water in the mix the density is slightly lower because the particle density of fly ash is lower (~2·3) than for PC (~3·12)
W/C ratio	0·50	0·42	In many countries, the fly ash is counted as being part of the cement. However, within the European standard (EN 206) fly ash is only partially counted as the cement using the 'k' factor route. The equivalent concrete performance route is an alternative approach

*The mix information is indicative of the changes that are likely to be required. Mixes should be checked by trial mixes in all casses.

of bleed can result in drying shrinkage cracking. The low heat properties and extended settings times of fly ash concrete lead to an increased risk of such cracking.

Special types of fly ash concrete

Roller-compacted concrete

The traditional method for the placing and compaction of concrete is for it to be laid in controlled layers between formwork. It is then compacted by an immersion vibrator to expel entrapped air. In contrast, roller-compacted concrete is a concrete mix with a high cementitious and low water content. This material is transported to site, often in open trucks, laid in discrete layers by earth-moving plant and consolidated by a vibrating roller. In this process the concrete, as compacted, has a low void ratio, high density and good bond between successive layers.

A development in the construction of concrete gravity dams saw a change from the construction of a series of blocks or monoliths separate from each other in the dam to the spreading of concrete in uniform layers over the whole length of dam compacted by a vibrating roller. This required the provision of temporary kerbs. From experience it had been found that a high paste content, cementitious material and water, was needed to bond successive layers of concrete together.

However, the relatively high cement content also generated heat of hydration extremes across the section, leading to shrinkage cracking and the development of internal strains in the concrete. Replacement of a proportion of the cement with fly ash would reduce these heat of hydration gradients.

High fly ash content roller-compacted concrete depends on achieving the optimum packing of all constituents in the concrete.[78] That means that all voids should be filled. It is usual practice to optimise the coarse aggregate content of a mix, but filling of the voids in the mortar fraction is rarely considered. For mix design the paste fraction is the absolute volume of the cementitious materials and free water. The mortar fraction is the absolute volume of the fine aggregate and the paste fraction.

A minimum paste content is necessary to fill the voids in the fine aggregate, while as low a cement content as possible is needed to reduce the heat of hydration. This offers the opportunity to add substantial quantities of fly ash, a material with the same particle size range as cement. The inclusion of fly ash will modify the rheology of the mix and reduce water demand. It is pozzolanic, will give long-term benefits and will lower internal temperatures in the hardening concrete. Adopting these principles has provided roller-compacted concrete with 60–80% by volume of fly ash being used in the mix.

High fly ash content concrete

Mix design principles developed for roller-compacted concrete have been used for concrete containing high volumes of fly ash and which may be compacted by immersion vibration plant.[79] Adopting the principle of minimum voids in the paste, mortar and aggregate, mix designs have been successfully used for concrete placed to floor slabs, structural basements and walls. These concretes have 40–60% of the cementitious volume as fly ash. Other work, using the maximum packing, minimum porosity principle,[80] have similarly been designed for structural concrete with fly ash making up to 70% of the cementitious content by weight.

Sprayed concrete with pulverised fuel ash

Sprayed concrete is a mixture of cement, aggregate and water, which may include fibres and/or admixtures, projected at high velocity from a nozzle into place to produce a dense homogeneous mass. It is sometimes called 'gunite' when the maximum aggregate size is <10 mm. 'Shotcrete' is the term for sprayed concrete where the maximum aggregate size is 10 mm or greater. Sprayed concrete may be applied by the 'dry' process, that is when the mixing water is added at the spray nozzle. The 'wet' process is a mixture of cement and aggregate weight batched and mixed with water prior to being conveyed through the delivery pipe to the nozzle.

The art of sprayed concrete largely depends on the experience and competence of the sprayed concrete nozzleman. However experienced the operator may be, the very velocity of placing the mixture causes some to bounce back or rebound from the surface under construction. This rebound is mainly the sand fraction, which may be as high as 30% for overhead sprayed concrete. Thus, there is considerable waste or rebound with the process.

Limited experimental work has shown that a major reduction in rebound can be achieved by replacing a proportion of cement with fly ash, or by using a blended fly ash–cement component. Using the 'dry' process of spraying and replacing 30% of the cement content with fly ash, a 30% reduction in generated rebound was recorded and at the same time the hardened concrete revealed a smaller pore size with no loss of compressive strength.[81] This finding has major potential for the replacement of a proportion of cement with PFA in sprayed concrete mixes.

Placing and compacting of fly ash concrete

The following information is provided to assist site engineers and foremen to achieve the best from concrete made with fly ash and illustrate how it can be used in all types of concrete structures.

Plastic properties of fly ash concrete

Fly ash is a fine material with a spherical particle shape. When added to concrete it produces a cohesive concrete, which looks drier than normal concrete of similar workability. Normally, less fine aggregate will be added to fly ash concrete in order to produce the best performance from the mix. The following factors should be taken into account.

- Fly ash concrete is often darker than Portland cement (CEM I)-only concrete. The colour consistency is similar to other concretes.
- Fly ash concrete visually appears more cohesive and less workable than CEM I-only concrete. Because of the rounded shape of fly ash particles, when vibrated the concrete will become highly mobile and will move readily within shutters. For this reason water should not be added on site to 'improve the workability' of the concrete based only on a visual assessment.
- Fly ash reduces the rate of bleeding within concrete. Bleed water that collects at the surface of concrete increases the W/C ratio and reduces the strength and durability. However, because less water rises with fly ash concretes they must be protected from excessive water loss, e.g. in drying windy weather conditions. If the surface of any concrete dries out before sufficient strength has developed, early age shrinkage cracking may occur. Protection and curing should be carried out at an early stage to prevent cracking problems.
- Fly ash concrete normally contains less water than the equivalent CEM I concrete. If used with a water-reducing admixture, e.g. plast ciser, the above effects are amplified. If retarding admixtures are used with concrete the risks of early age drying shrinkage cracking is increased.
- The lower the water content of the concrete the less effective is the poker vibrator at compacting the concrete, irrespective of the cementitious type. Because fly ash concrete has a lower water content, the poker should be placed at closer centres for a longer period to ensure full compaction of such mixes.
- Self-compacting concrete (SCC) should be considered. This needs no vibration to compact the concrete.

Achieving the best results from the concrete

As stated above, fly ash is a pozzolana that reacts with the lime produced when CEM I is mixed with water. The pozzolanic reaction is temperature dependent. The following should be considered after the concrete is placed.

- Fly ash reduces the amount of heat produced in comparison with Portland cement concretes of the same strength. In thick sections this is a benefit and reduces the risk of thermal cracking. However, for thin sections excessive heat loss may reduce strength, e.g. in cold weather conditions. In cool or cold weather conditions, concrete should be protected from heat loss, both in the structure and for test cubes. Test cubes should be moved to a heated room at 20 ± 5°C after casting, and the concrete insulated to keep the chemical reactions going.

- If stripping shutters are removed too early, or if the surface of the concrete has been allowed to cool too much, a weak friable layer of concrete may lead to scabbing. One should refer to the specification for the minimum stripping times, e.g. BS 8110, Highways Specification.
- In hot and drying weather conditions, the importance of proper curing regimes cannot be overemphasised. Any exposed surface should not be allowed to dry out for at least 3 days, and preferably longer. If kept wet and warm for long periods fly ash concrete can produce highly durable concrete second to none.
- Test cubes must be stripped after 24 h and stored under water at 20 ± 5°C, or preferably at 20 ± 2°C. Cubes must be fully compacted using a representative sample of the concrete taken throughout the discharge of the load. One should take care not to use excessive amounts of mould oil in cube moulds and ensure that the moulds are in good condition: they should be checked annually. Cubes should be labelled so that errors cannot occur but this should not be done by scratching the surface of the cube.

Hardened concrete

Fly ash concrete is very similar in most respects to Portland cement concrete. The following are a few factors to consider.

- *Colour*: During the first few days and weeks after casting, concrete changes in colour as hydration of the cement proceeds. As fly ash acts as a pozzolana, these reactions will continue for many years in the presence of water. However, as fly ash uses excess lime there is a reduced risk of efflorescence from the concrete. Colour is also affected by the absorption of water by formwork materials. Absorbent form face materials tend to produce dark concrete. With increasing reuse of formwork, lighter concrete will appear as the absorption of the form face reduces because the pores are blocked with fine particles and mould oils.
- *Protection*: All concrete should be protected from physical damage. The strength development of all concrete takes time and arrises are easily damaged. Access to the area should be prevented or protection provided to exposed edges, corners, etc.
- *Making good*: No special procedures are required for fly ash concrete. All 'making good' operations should be avoided wherever possible; these is no substitute for care in construction.

- *Durability*: With proper site practice fly ash concrete can be exceptionally durable. The Romans used pozzolanic materials like fly ash to build the Pantheon in AD 115. There is no reason why concrete on a modern site should not last over 1000 years.
- *Personal protective equipment*: PPE appropriate for normal concreting operations should be worn. Concrete has a high pH value and should be handled with care.
- *Environmental*: Specifying the use of fly ash in concrete benefits the environment by replacing manufactured materials with industrial by-products. As concrete has a high pH value, it should be disposed of appropriately.

Examples of fly ash concrete in the UK

Pozzolanic materials have a long history of producing highly durable concretes. The following are UK structures, which were constructed using unclassified fly ash prior to the inception of controlled fineness with British Standards, e.g. BS 3892 Part 1.

- Lednock dam[6] was the first significant construction within the UK that used fly ash.
- Ferrybridge power station,[82] Yorkshire: 16% replacement of cement with fly ash in the foundations where over 49,000 m^3 of concrete was used.
- Stithians dam[83] near Redruth, Cornwall: 25% and 30% replacement of cement was used in the construction of this concrete dam; some 36,000 m^3 of concrete was used.
- Pembroke Power Station:[84] 104,000 m^3 of concrete was used in the construction of culverts and other below-ground structures.
- Leith harbour development,[85] east coast of Scotland: 180,000 m^3 of concrete was used in hearting and facing construction; 25% fly ash replacement of cement was used. The standard deviation was 3·1 MPa on a grade 16·5 MPa mix.
- Ragdale terminal reservoir,[86] north of Leicester: 30% fly ash was used in the construction of the mass concrete water-retaining perimeter wall, the lower floor and blinding layers.
- Upper Tamar reservoir,[87] near Bude, North Cornwall: 20–27% fly ash was used in the concrete mixes for the spillways, facing and heart concrete within the dam.
- Drax power station[88] used selected fly ash with LOI up to 12%.

Fig. 3.24. Dinorwig power station tunnels, UK

Many more structures have been built with controlled fineness PFA to BS 3892 Part 1, including:

- Dinorwic pumped storage scheme (Fig. 3.24):[89] this involved the production of concrete-lined shafts used as pumped water storage conduits for electricity production
- the Thames Barrier (Fig. 3.25)
- the London Docklands development
- Maplethorpe sea defence works
- Sizewell B nuclear power station
- The Channel Tunnel.

Summary

The benefits of using fly ash in concrete can be summarised as follows.

1. Lower water contents at the same workability with consequent reductions in bleeding, drying shrinkage and permeability.
2. Better placeability, increased fine solids content and higher-quality surface finishes.

Fig. 3.25. The Thames Barrier, UK

3. Lower heat of hydration reducing the potential for thermal cracking problems.
4. Longer-term strength gain properties. With Portland cements, high initial curing temperatures will cause a rapid initial strength gain but a relatively reduced 28 day strength. With concrete containing fly ash, at normal curing temperatures, in relation to Portland cement concrete, lower early strengths are obtained but at later ages, e.g. $\geqslant$ 56 days, superior strengths are found.
5. Improved durability, because fixation of the lime produced during the Portland cement hydration reduces the permeability of fly ash concretes. This increases the long-term strength, as in Fig. 3.15,[90] and lowers the susceptibility to sulfate attack, chloride ingress, acid attack, etc.
6. Environmental and financial cost savings: if the destination is within reasonable transportation distances of the supplying power station, the cost of fly ash is normally less than Portland cement. Significant cost savings are possible with careful mix design and by

fully utilising the long-term strength properties. Fly ash reduces the Portland cement content of a mix. As the Portland cement process involves the calcining of limestone its manufacture involves high-energy input and is a significant contributor to CO_2 emissions. With coal-burning power stations the environmental burden is directly associated with the production of electricity. By the use of fly ash, the overall CO_2 burden and energy consumption are reduced as a result of the reduction of the Portland cement component.

The use of fly ash in concrete is beneficial to the performance and durability of a structure in most applications. However, both concrete and fly ash have been researched to such an extent that at times it becomes difficult to see these benefits because of the extensive piles of papers. It is expected that the ever-increasing environmental pressure on industry will promote the further use of materials such as fly ash. Perversely, the same environmental pressures are reducing the amount of coal burnt, reducing the availability of fly ash of suitable quality. It is expected that recovery of older ash stockpiles will become economic to compensate for the reduced supply.

Whether the drying and reprocessing of stockpiled ash is environmentally beneficial is a matter for some conjecture. However, Dhir[91] has shown that conditioned ash can be used in concrete successfully. Perhaps when the handling problems of moist fly ash have been resolved there will be no need for dry fly ash to be produced.

References

1. Stanley CC. *Highlights in the history of concrete*. British Cement Association, Crowthorne, 1979. ISBN 0-7210-1156X.
2. Rivera-Villarreal R, Cabrera JG. The microstructure of two-thousand year old lightweight concrete. International Conference, Gramado, Brazil, 1999.
3. McMillan FR, Powers TC. A method of evaluating admixtures. *Proceedings of the American Concrete Institute* 1934; **30**: 325–344.
4. Davis RE, Kelly JW, Troxell GE, Davis HE. Proportions of mortars and concretes containing Portland–pozzolan cements. *ACI Journal* 1935; **32**: 80–114.
5. Davis RE, Carlston RW, Kelly JW, Davis HE. Properties of cements and concretes containing fly ash. *ACI Journal* 1937; **33**: 577–612.
6. Fulton AA, Marshall WT. The use of fly ash and similar materials in concrete. *Proceedings of the Institute of Civil Engineers Part 1* 1956; **5**: 714–730.
7. Allen AC. Features of Lednock Dam, including the use of fly ash. Paper No. 6326. *Proceedings of the Institute of Civil Engineers* 1958;**13**(August): 179–196.
8. Richardson L, Bailey JC. *Design, construction and testing of pulverised fuel ash concrete structures at Newman Spinney Power Station, Parts I, II, III*. CEGB Research and Development Report, 1965.

9. Howell LH. *Report on pulverised fuel ash as a partial replacement for cement in normal works concrete*. CEGB, East Midlands Division, 1958; also Ashtech '84, London, 1984.
10. Central Electricity Generating Board. Technical Bulletins Nos 1 to 48. CEGB, London.
11. BS 3892. *Pulverised fuel ash for use in concrete*. BSI, London, 1965.
12. *Pozzolan – a selected fly ash for use in concrete*. The Agrèment Board Certificate No. 75/283.
13. BS 3892. Part 1: *Pulverised fuel ash for use as a cementitious component in structural concrete*. BSI, London, 1982.
14. BS EN 450. *Fly ash for concrete – definitions, requirements and quality control*. BSI, London, 1995. ISBN 0-580-24612-4.
15. BS 3892. Part 1: *Specification for pulverised-fuel ash for use with Portland cement*. BSI, London, 1997. ISBN 0-580-26785-7.
16. BS EN 206. Concrete – Part 1: Specification, performance, production and conformity. BSI, London, 2001.
17. Smith IA. *The design of fly ash concretes*. ICE, Paper 6982, 1967: 769–790.
18. Dhir RK, McCarthy M, Magee BJ. *Use of fly ash to EN450 in structural concrete*. University of Dundee, Research Project Report, March 2000.
19. BS 8500 (Draft for public comment). *Concrete – complementary British Standard to BS EN 206-1*. BSI, London, 2001.
20. PrEN 12620. *Aggregates for concrete including those for use in roads and pavements*. Draft EN, 1996.
21. BS EN 12878. *Pigments for colouring of building materials based on cement and/or lime – Specifications and methods of test*. BSI, London, 1999.
22. PrEN 13263. *Silica fume for concrete – Definitions, requirements and quality control*, Draft EN, 1998.
23. BS EN 197-1. *Cement – Part 1: Composition, specifications and conformity criteria for common cements*. BSI, London, 2000.
24. BS 8500. *Concrete - complementary standard to BS EN 206 -1*, Parts 1 to 4, 2000 (draft).
25. BS EN 196-1. *Methods of testing cement, determination of strength*. BSI, London. 1995.
26. BS 3892. Part 2: *Specification for pulverised-fuel ash for use as a type I addition*. BSI, London, 1996. ISBN 0-580-26444-0.
27. BS 3892. Part 3: *Specification for pulverised-fuel ash for use in cementitious grouts*. BSI, London, 1997. ISBN 0-580-27689-9.
28. BS EN 450. *Fly ash for concrete – definitions, requirements and quality control*. BSI, London, 1995. ISBN 0-580-24612-4.
29. NUSTONE Environmental Trust. *The effect of the fineness of fly ash (fly ash) on the consistence and strength properties of standard mortar*. Project Report, March 2000. Read in conjunction with: EN 450 *Fly Ash* and BS 3892. *Part 1: Testing Program*. UKQAA, Analysis of results, March 2000.
30. BS EN 196-5. *Methods of testing cement, pozzolanicity test for pozzolanic cements*. BSI, London, 1995, ISBN 0-580-21525-3.
31. BS 5328. Part 1: *Concrete – guide to specifying concrete*. BSI, London, 1997. ISBN 0-580-26722-9.

BS 5328. Part 2: *Concrete – methods for specifying concrete mixes*. BSI, London, 1997. ISBN 0-580-26723-7.
BS 5328. Part 3: *Concrete – specification for the procedures to be used in producing and transporting concrete*. BSI, London, 1990. ISBN 0-580-18979-1.
BS 5328. Part 4: *Concrete – specification for the procedures to be used in sampling, testing and assessing compliance of concrete*. BSI, London, 1990. ISBN 0-580-18980-5.

32. BS 6610. *Specification for pozzolanic pulverised fuel ash cement*. BSI, London, 1996. ISBN 0-58025411-9.
33. Roy DM. *Hydration of blended cements containing slag, fly ash or silica fume*. Sir Frederick Lea Memorial Lecture, 29 April–1 May 1987, Institute of Concrete Technology Annual Symposium, 1987.
34. Dhir RK. Pulverised fuel ash. *CEGB Ash-Tech 86 Conference Proceedings*, 1986.
35. Cabrera JG, Plowman C. Hydration and microstructure of high fly ash content concrete. *Conference on concrete dams*, London, 1981.
36. Cabrera JG, Gray MN. Specific surface, pozzolanic activity and composition of pulverised fuel ash. *Fuel Magazine* 1973; **52:** 213–219.
37. Thomas MDA, Mathews JD, Haynes CA. The effect of curing on the strength and permeability of fly ash concrete. *ACI Journal* 1989; **SP-114:** 191–217.
38. Cabrera JG, Plowman C. Hydration and microstructure of high fly ash content concrete. *CIRIA, Conference on Concrete Dams*, London, 1981.
39. Taylor HFW. *Cement Chemistry*, 2nd Edn. Thomas Telford Publishing, London, 1997. ISBN 0-7277-2592-0.
40. Owens PL. Fly ash and its usage in concrete. *Concrete Magazine* 1979; July, 22–26.
41. Monk MG. Portland-fly ash cement: a comparison between intergrinding and blending. *Magazine of Concrete Research* 1983; **35**(124)**:** 131–141.
42. Paya J, Monzo J, Borrachero MV. Mechanical treatment of fly ashes, Part II. *Cement and Concrete Research* 1996; **26**(2)**:** 225–235.
43. Dhir RK, Munday JGL, Ong LT. Strength variability of OPC/fly ash concrete. *Concrete Magazine* 1981; June, 33–38.
44. Matthews JD, Gutt WH. Studies of fly ash as a cementitious material. *Conference on Ash Technology and Marketing*. London, October 1978.
45. Cabrera JG, Hopkins CJ. *The Drax project – the properties of pulverised fuel ashes and their relevance to the properties of fly ash concrete*. Research Report, 1986.
46. Bogdanovic M. The use of fly ash in the concrete of a curbstone of a railway bridge. *Conference on Ash Technology and Marketing*, October 1978.
47. Bouzoubaa N, Zhang M-H, Malhotra VM. Blended fly ash cement – a review. *ACI International Conference on High Performance Concrete*, Kuala Lumpur, 2–5 December, 1997, pp. 641–650.
48. Dewar JD. *The particle structure of fresh concrete – a new solution to an old question*. Sir Frederick Lea Memorial Lecture, Institute of Concrete Technology Annual Symposium, 1986.
49. Brown JH. *The effect of two different pulverised fuel ashes upon the workability and strength of concrete*. C&CA Technical Report No. 536, June 1980.

50. Concrete Society Core Project. *Data from potential revision of Technical Report No. 11*, 1998.
51. Woolley GR, Conlin RM. Pulverised fuel ash in construction of natural draught cooling towers. *Proceedings of the Institute of Civil Engineers, Part 1*. Paper 9278, 1989; **86:** 59–90.
52. Bamforth PB. Heat of hydration of fly ash concrete and its effect on strength development, *Ashtech '84 Conference*, London, 1984: 287–294.
53. Woolley GR, Cabrera JG. Early age in-situ strength development of fly ash concrete in thin shells. *International Conference on Blended Cement*, Sheffield, 1991.
54. BS 8110. *Structural use of concrete*. Code of practice for design and construction. BSI, London.
55. Dhir RK, Munday JGL, Ong LT. Investigations of the engineering properties of OPC/pulversied fuel ash concrete – deformation properties. *Structural Engineer* 1986; **64B**(2)**:** 36–42.
56. Browne RD. Ash concrete – its engineering performance. *AshTech '84*, London, 1984: 295–301.
57. Gifford PM, Ward MA. Results of laboratory test on lean mass concrete utilising fly ash to a high level of cement replacement. *Proceedings of an International Symposium*, Leeds, 1982: 221–229.
58. Keck RH, Riggs EH. Specifying fly ash for durable concrete. *Concrete International*, 1997; April, 35–38.
59. Thomas MDA. *A comparison of the properties of OPC and fly ash concrete in 30 year old mass concrete structures, Durability of building materials and components*. E & F N Spon, London, 1990: 383–394.
60. Berry EE, Malhotra VM. Fly ash in concrete. CANMET, SP85-3.
61. Concrete Society. *Permeability of concrete*. Concrete Society, London, 1985: 6–68.
62. Concrete Society. *Alkali silica reaction: minimising the risk of damage*. Technical Report No. 30, 3rd edn. Concrete Society, London, 1999.
63. BS 5328. *Part 2: 1997 Methods for specifying concrete, amendment 10365* BSI, London, May 1999.
64. Alasali MM, Malhotra VM. Role of concrete incorporating high volumes of fly ash in controlling expansion due to alkali–aggregate reaction, *ACI Materials Journal* 1991; **88**(2)**:** 159–163.
65. Taylor HFW. *Cement chemistry*. Thomas Telford Publishing, London, Reprint 1998. ISBN 0-7277-2592-0.
66. *ACI manual of concrete practice, Fly ash*, 226.3R, 1994.
67. Fournier B, Malhotra VM. CANMET investigations on the effectiveness of fly ash in reducing expansion due to alkali aggregate reaction (ASR). *ACAA 12th International Symposium*, 1997.
68. Concrete Society. *The use of GGBS and fly ash in concrete*. Technical Report No. 40. Concrete Society, Crowthorne, Berks, 1991. ISBN 0-946691-40-1.
69. Cabrera JG, Woolley GR. A study of twenty five year old pulverised fuel ash concrete used in foundations. *Proceedings of the Institute of Civil Engineers* 1985; **79**: 149–165.
70. Dhir RK, Munday JGL, Ho NY. *Fly ash in concrete: freeze thaw durability*. Draft report. University of Dundee, 1987.

71. Sulfate and acid resistance of concrete in the ground. *BRE Digest* 1991; **363**, (July).
72. Bensted J. Thaumasite – a deterioration product of hardened cement structures, Il. *Cemento magazine* 1988: 3–10.
73. Burton MW. *The sulphate resistance of concretes made with ordinary Portland cement, sulphate resisting cement and ordinary Portland cement + pozzolan.* Kirton Concrete Services, Humberside, 1980.
74. Jackson AJW, Goodridge WF. A new approach to P.F. ash concrete. *Contract Journal* 1961; **180:** 1284–1296.
75. Sear LKA. The development of a decision support system for the quality control of readymixed concrete. PhD Thesis, University of Wolverhampton, August 1996.
76. Dewar J. *The Particle Structure of Fresh Concrete – a new solution to an old question.* Sir Frederick Lea Memorial Lecture, Institute of Concrete Technology, 1986.
    Department of the Environment. *Design of normal concrete mixes.* BRE, Watford.
    Owens PL. *Basic mix design,* CEGB Ash-Tech Conference Proceedings C74, pp. 29–35, 1974.
77. Hobbs DW. Portland-pulverized fuel-ash concretes: water demand, 28 day strength, mix design and strength development. *Proceedings of the Institute of Civil Engineers*, Part 2, Paper 9322. 1988; **33:** 317–331.
78. Dunstan MRH. *Rolled concrete for dams.* CIRIA Technical Note No. 106. CIRIA, London, 1981.
79. Cabrera JG, Atis CD. *Design and properties of high volume fly ash performance concrete,* ACI Proceedings, SP 196, Michigan, 1998.
80. Cabrera JG, Braim M, Rawcliffe J. *The use of pulverised fuel ash for construction of structural fill.* AshTech, London, 1984.
81. Cabrera JG, Woolley GR. Properties of sprayed concrete containing ordinary Portland cement or fly ash Portland cement. Proceedings of the ACI/SCA International Conference, Edinburgh, 1996: 8–24.
82. *Ferrybridge C power station.* CEGB datasheet No. 1. CEGB, London, 1965.
83. *Stithians dam.* CEGB datasheet No. 2. CEGB, London, 1965.
84. *Pumping concrete at Pembroke power station.* CEGB datasheet No. 28. CEGB, London, 1969.
85. *Leith harbour development.* CEGB datasheet No. 31. CEGB, London, 1979.
86. *Ragdale terminal reservoir, Leicester.* CEGB datasheet No. 39. CEGB, London, 1972.
87. *The upper Tamar reservoir.* CEGB datasheet No. 41. CEGB, London, 1974.
88. *Fly ash concrete for Drax power station completion works.* CEGB technical bulletin No. 5. CEGB, London, 1985.
89 Copeland BGT. Fly ash concrete for hydraulic tunnels and shafts, Dinorwic pumped storage scheme – case history. *Conference proceedings,* 1981.
90. *In situ concrete strength: an investigation of the relationship between core strength and standard cube strength.* Concrete Society Project Report No. 1, The Concrete Society, 2000.
91. Dhir RK. *The use of conditioned ash in concrete.* University of Dundee, 1999.

Chapter 4

Fly ash as a fill material

Introduction

Fly ash has been successfully used as a fill for many years, with the first recorded use in the UK being in 1952. A considerable amount of research was done in the 1950s and 1960s, which formed the basis of its use. It is acknowledged to have benefits of low density and high shear strength, which have been instrumental in developing its wide acceptance as a fill material.

Types of fly ash

For the purposes of fill, fly ash can be considered as being available in three forms: conditioned, stockpile and lagoon (Fig. 4.1). Chapter 1 describes the production, storage and properties of these types of fly ash.

Properties of fly ash as a fill material

As described in Chapter 1, fly ash will gain strength with the passage of time. Work at Newcastle University,[1] following earlier work at Glasgow and Salford, looked at the age hardening of fly ash and found a number of factors influencing the strength. When compacted, suctions develop in the fly ash which result in a cohesive force in the fly ash. These will dissipate slowly but disappear if the material becomes saturated. After a short period there is growth of gypsum crystals that creates bonding between the fly ash particles. This also results in a rougher surface to the particles, increasing the friction angle. The resulting strength is not lost if the fly ash becomes saturated, depending on the gypsum content.

If sufficient lime is present in the fly ash then it will result in further hardening due to a combination of further crystal formation and reaction between the calcium oxide and the glassy material in the fly ash

Fig. 4.1. A fly ash lagoon

(pozzolanic reaction). It has been noticed[2] that it is not the total calcium content that is important but the free calcium oxide, that is the amount that can be brought into solution and is available to react. An increase in free calcium oxide will result in greater strength gain with time. However, there does not appear to be a simple relationship between total calcium content and the free calcium oxide and therefore it is not easy to predict the strength gain of a fly ash.

When water is added to fly ash, it initially has a low pH[3] as the sulfate deposited on the surface of the particles is brought into solution as sulfuric acid. This is a transient situation and the pH rapidly rises as calcium is leached into solution (Fig. 4.2). The pH is typically 9–11 for fly ash, although the pH for those ashes with higher free calcium oxide contents can rise to 12. Only a very small quantity of free calcium is required to achieve the higher pH. Because most of the water-soluble material that influences pH has been washed out of lagoon fly ash, the pH is lower, typically around 9.

The calcium content of fly ash means that most of the sulfate is present as gypsum, which has a limited solubility. When tested in accordance with BS 1377[4] the water-soluble sulfate content of conditioned and stockpile fly ash is typically 2·0–2·5 g/l, which means that it is on the boundary of sulfate classes 2 and 3 as defined in BRE Digest 363.[5] However, the permeability of

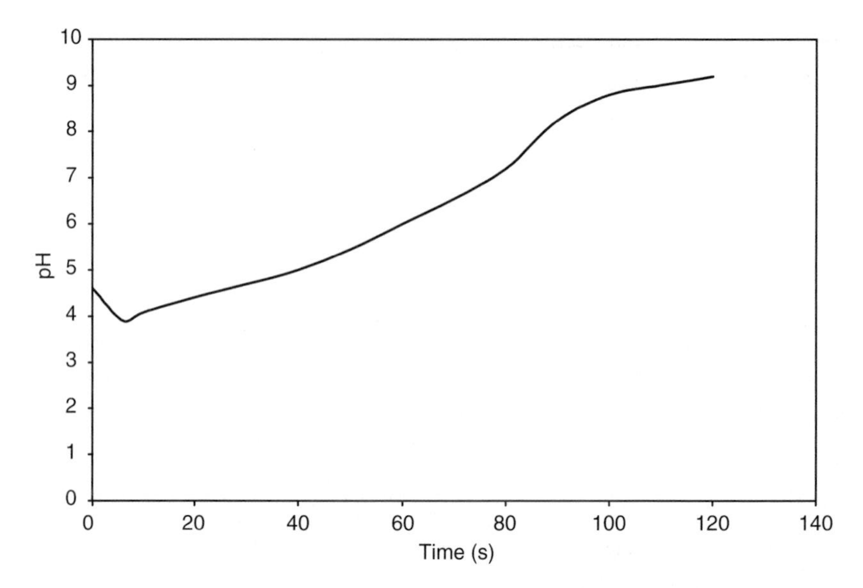

Fig. 4.2. When water is added to fly ash it quickly changes from being acidic to alkaline

fly ash is well below 10^{-5} m/s, which means that there will be restricted movement of groundwater through the fly ash, so class 2 exposure conditions are the most appropriate classification. The sulfate level of lagoon fly ash is usually very low, as the majority of the sulfate will have been washed out. The sulfate content is typically $<0{\cdot}1$ g/l, so exposure class 1 is more appropriate.

The sulfate content of fly ash means that it cannot be placed within 500 mm of metallic items, according to the Department of Transport Specification for Highway Works (SHW). The water-soluble content of fly ash is also sufficiently high to restrict the types of reinforcement that can be used in reinforced earth structures. This is discussed in more detail later in this chapter.

The loss on ignition (LOI) is a measure of the carbon content. The carbon has a low density and can absorb significant amounts of water. This means that the maximum dry density and optimum moisture content of fly ash are influenced by the LOI. Higher LOI ashes are lower in density, but have higher optimum moisture contents.

Density

The particle density of fly ash is typically 2·0–2·4 mg/m^3, the lower density being associated with a high LOI. There is some variability in the density of particles, with smaller ones having higher densities. This is due to air voids within many of the particles, and between 1% and 5% contain sufficiently large voids that they float on water. The variation in particle density means that sedimentation techniques for determining the particle size distribution are not suitable and more appropriate methods are now used, e.g. laser scattering.

Over the years, it has been accepted that the most appropriate method for determining the compaction parameters in the laboratory is using the 2·5 kg rammer, as detailed in BS 1377.[6] The heavy compaction will give higher maximum dry densities and lower optimum moisture content values, but they are only slightly different from the light compaction; the latter produces more realistic target values for site control.

Typical compaction data are shown in Table 4.1. The compaction data vary from station to station, so it is important that the source of fly ash is established and that the data for that source are obtained. Typical curves for a range of sources are shown in Fig. 4.3. The variation in maximum

Table 4.1. Typical data for compacted fly ash

Parameter	Range
Bulk density	1·5 mg/m^3–1·8 g/m^3
Optimum moisture content	14–35%
Maximum oven-dry density	1·1–1·6 mg/m^3

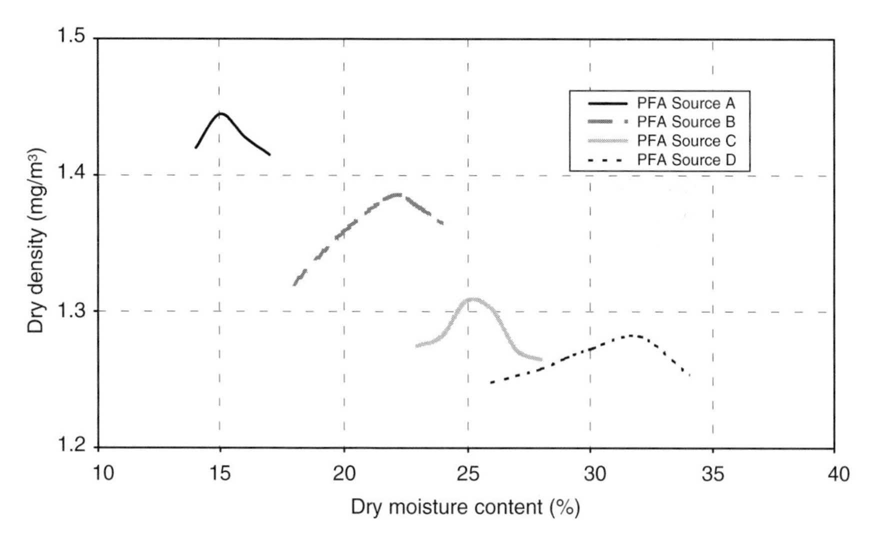

Fig. 4.3. Typical compaction curves, showing that optimum moisture content varies with pulverised fly ash (PFA)/fly ash source

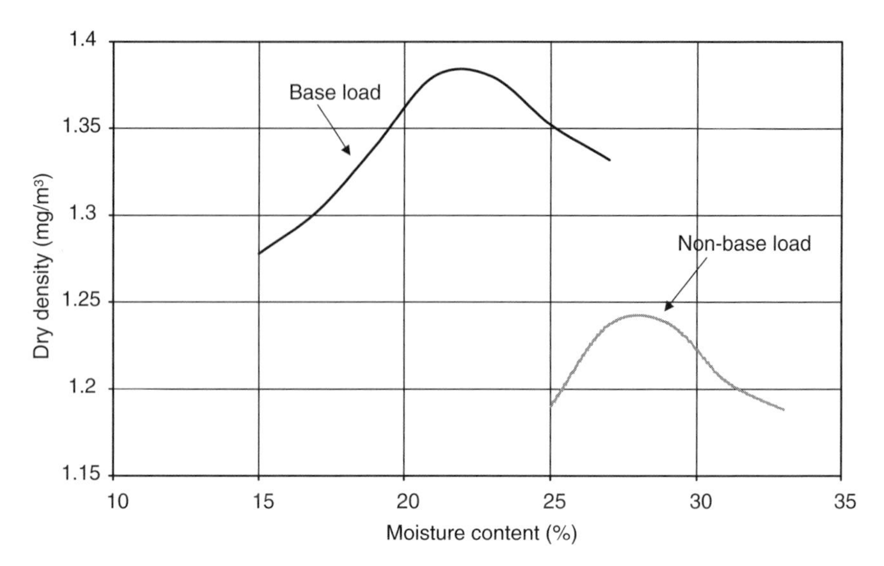

Fig. 4.4. Effect on density and optimum moisture content of power station loading

dry density and optimum moisture content can be seen, with denser ashes having lower optimum moisture contents. The curves for the ashes with lower optimum moisture contents tend to have more pronounced peaks and consequently the density is more sensitive to changes in moisture content.

As well as variation in compaction data between sources, there can be variations within a source. The plot in Fig. 4.4 indicates the changes that can occur when the load factor at a power station changes. When a station runs continuously (base load) the loss on ignition of the fly ash will be low, resulting in a high-density and low optimum moisture content. Conversely, when the power station has a fluctuating load the LOI will rise, affecting the density and optimum moisture content. However, it should be noted that some sources have low-density fly ash with low LOI.

The air void content of compacted fly ash is relatively high. At maximum dry density, the fly ash usually contains between 5% and 10% air voids (Fig. 4.5), and some can even lie close to 15% air voids. This means that air voids are not a reliable way of measuring the compaction of fly ash.

Compaction

Trials in the 1960s carried out by the Central Electricity Generating Board (CEGB), Lancashire County Council and Stevenson Clarke and reported by Smith[7] laid the basis of placing and compacting fly ash as a fill. The general guidelines are to place the fly ash in 200–225 mm loose and

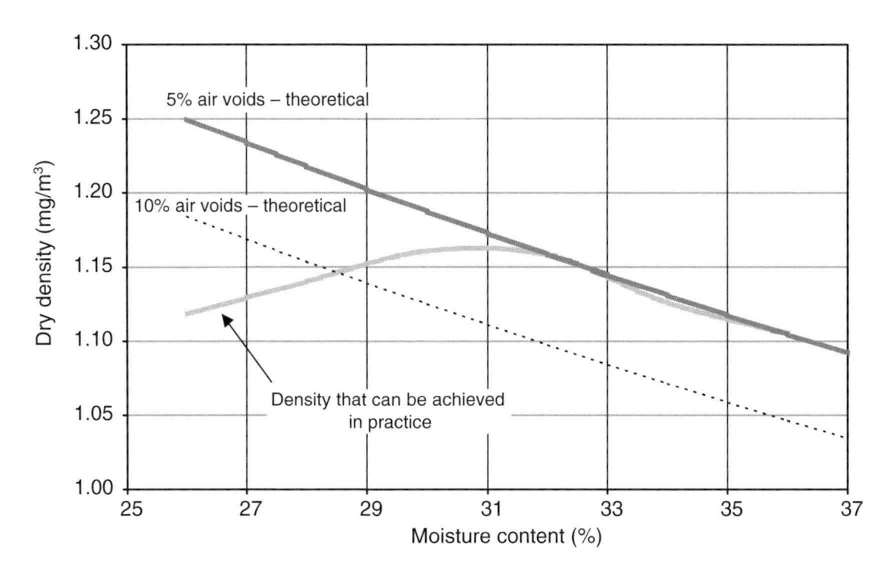

Fig. 4.5. Typical air voids plot

compact layers with eight passes of a roller. It is possible to place fly ash in thicker layers and successfully compact it but care needs to be taken. If the material is drier than the optimum moisture content then the compactive effort is dissipated in the top of the layer and the bottom may not be properly compacted.

It is usual in the UK to use vibrating rollers to compact fly ash because these are widely available. Rollers towed by a tracked bulldozer give good results, with the tracks of the dozer assisting in the compaction. Although vibrating rollers have been shown to give acceptable results, they tend to overstress the top 50 mm of the fly ash. This will heal when the next layer is placed on top or can be sealed by running the roller over without vibration for the final pass. Smith noticed that fly ash was compacted by lorries running over it. Consequently, pneumatic-tyred rollers were included in the compaction trials. These were shown to be very effective in compacting fly ash, producing a good, close-knit surface. These types of roller are used widely elsewhere, including the USA.[8] The recommendations are to use self-propelled rollers with a dead weight of 10–12 tonnes and tyre pressures not exceeding 250 kPa.

Sheepsfoot rollers, smooth-wheeled rollers and grid rollers have been found to be unsuitable and vibrating plates should only be used in small areas where access is difficult and layer thickness can be carefully controlled.

Cotton[9] reports that pneumatic-tyred rollers were used on several contracts on the M6 in Lancashire and Cheshire in the early 1960s. The lack of availability, cost and the fact that they were not accepted as suitable for compaction of fly ash in the 1976 version of SHW, UK, restricted their use. A pneumatic-tyred roller was successfully used in the construction of the G-Mex Centre in Manchester. The area was supported by a series of arches and a lightweight fill was required, but there was concern that vibrating rollers might weaken the structure. A pneumatic-tyred roller normally used on surface dressing work was employed, with concrete blocks added to provide ballast, and this achieved the desired level of compaction of the fly ash fill.

The SHW requires fly ash to be compacted to 95% of maximum dry density. This has proved a difficult requirement to meet. The fly ash from a single base load source is usually consistent, with a variation of less than 3% in maximum dry density measured over time. This is still sufficiently variable to cause problems if a single target maximum is used. The Road Research Laboratory in 1966[10] reported that a method specification was more appropriate for control of the density of fly ash. It was also noted that there was no need to monitor the moisture content of the fly ash as this appeared to have little effect on the density achieved. This may be due to the particular ashes having relatively flat curves. It is more common to have trouble with compaction if the moisture content of the fly ash is low. Fly ash can dry out rapidly, especially in warm, windy conditions and it is therefore recommended that water is available on site to add to the fly ash to overcome problems associated with drying.

If a method specification is to be used, then control of the quality of the fly ash on delivery to site is required. The moisture content can be monitored and 95% compaction can be achieved if the moisture content is maintained within the range 0·8–1·2 times the optimum. However, if the optimum moisture content is very variable, this may not be a suitable means of control. The moisture condition value apparatus has been investigated as a control tool for fly ash suitability, but opinions as to its suitability are varied. Reducing the drop height may produce results that are more meaningful. Once an understanding of the nature of fly ash has been gained, it is possible to assess the suitability of the fly ash by squeezing a pat in the hand. If it binds together then it is considered acceptable.

Density is usually measured in the penultimate layer to avoid any influence of overshearing. Core-cutter and sand-replacement methods have both been found to be acceptable, core-cutter especially so because of the speed. Nuclear density meters are also suitable for bulk density measurements, provided the fly ash has a consistent LOI, as the carbon can influence the results. Microwave ovens can be used to give a rapid measure of moisture content, but care has to be exercised because excessive heating can lead to the carbon burning off with high carbon ashes. When drying lagoon or stockpile fly ash in a microwave oven the sample container should be covered because there is a risk of the samples' disintegrating and being lost. The 'Speedy' moisture meter has also proved successful as a means of measuring moisture content.

Permeability and capillarity

The permeability of fly ash is relatively low and values have been quoted[11] to be in the range $0{\cdot}01 \times 10^{-7}$ and 8×10^{-7} m/s, meaning that it has poor drainage characteristics. Experience has shown that if fly ash is well compacted and is subsequently subjected to heavy rain, it will slowly absorb moisture, the top surface may become saturated and the majority of the rain will be shed. There will only be slow penetration of water into the fly ash and studies from several disposal sites have indicated that there is no conclusive evidence of percolation through the mounds.

If the upper surface of fly ash becomes saturated then it will recover rapidly once the weather becomes drier. If necessary, the wet material can be removed, stored and re-used when it has dried out. Alternatively, semi-dry fly ash can be added to absorb moisture.

It is important to protect the side slopes of embankments as soon as possible after completion, usually with topsoil, to prevent channels being scoured out of the fly ash. Care must be taken to prevent excessive run-off during the construction stage, wherever possible.

When the base of compacted fly ash becomes saturated, the water will be drawn up by capillary action to 0·5–0·6 m above the water level. This can cause instability problems in thin layers, $<$600 mm thick. It is recommended that to avoid such problems, a drainage blanket should be placed under the fly ash. This should be sufficiently thick to raise the fly ash above the water level; a drainage blanket of sand with a thickness of 300–450 mm is usually recommended. As well as sand, crushed rock, crushed concrete and slag are all acceptable as a drainage material, providing the grading is correct. There have been instances where fly ash has been successfully placed in wet areas, one example being the embankments constructed as part of the Oakham Ness oil terminal.[12] Here, two fly ash embankments were built on saltings; the first layer was fly ash end-tipped on to the marshy area to a depth of 600 mm without compaction. Subsequent layers were spread and rolled with a 10 tonne pneumatic-tyred roller. The side slopes were faced with stone to prevent erosion. Another interesting application was reclamation of land adjacent to the River Medway at Lappel Bank,[13] Sheerness. Here, lagoon fly ash was pumped through 1000 m of pipework into barges. The barges transported the fly ash down river to the Lappel Bank reclamation site.

Because fly ash is a silt-like material, it can be considered to be susceptible to frost. Work done by the Road Research Laboratory[14] on seven different ashes found that four of them were frost susceptible. They concluded that ashes with more than 40% retained on the 75 μm sieve were susceptible, but these are exceptionally coarse ashes. The frost susceptibility of fly ash was confirmed in work reported by Sutherland and Gaskin.[15] Because of the potential for frost heave, it is recommended that fly ash is kept at least 450 mm below the finished surface.

Settlement

Fly ash has a stiffness similar to a hard clay, with M_v typically 0·1–0·2 MN/m^2, depending on the degree of compaction. Values for the elastic modulus, based on secant measurements from undrained triaxial tests, at 1·67% strain, have been quoted by Barber *et al.*[11] as typically 70 MN/m^2 for tests carried out immediately after compaction, rising to around 110 MN/m^2 at 28 days.

Raymond and Smith[16] noted that settlements predicted from laboratory and, to a lesser extent, field test results overestimated settlements. The Thermalite factory at Agecroft in Manchester was built on 15 m of stockpile fly ash that had not had any systematic compaction. The settlements calculated from laboratory consolidation tests were found to be in the range of 300–405 mm. Standard penetration tests* gave an average *N*-value of 8·5. It was estimated from this that settlements would be in the order of 25 mm. Plate-bearing tests were then carried out on the site, from which it was estimated that settlement would be in the range of 0·5–18·5 mm. Actual measurements showed the settlement to be, in general, less than predicted by the plate-bearing test.

Sutherland *et al.*[2] reported the use of mixed lagoon fly ash in filling a 6 m deep railway cutting on the A452 at Packington in 1952. The material, with a moisture content of 55%, was end-tipped into the cutting and given no compaction. Despite the method of placing, the fly ash only settled around 38 mm in 2 months while it was temporarily carrying the traffic. The permanent road was then constructed and after 4 years no further settlement had been detected.

Cabrera *et al.*[17] examined the effect of applying load to a compacted lagoon fly ash in a field trial. An area 10·5 × 8·5 m × 1·5 m deep was excavated and filled with fly ash under controlled conditions. A reinforced concrete raft 4·7 × 2·5 m × 0·15 m thick with a down-stand 0·3 m square-section edge beam was cast on the surface of the fill. The raft was evenly loaded with a surcharge of 55 tonnes. Settlement was measured at nine reference points on the surface of the raft. The measurements showed that

*The standard penetration test involves driving a sampler using a standard rammer 450 mm into the soil. The number of blows required to achieve the last 300 mm of penetration is the *N*-value. It is used to estimate the relative density, bearing capacity and friction angle of the soil.

the majority of the settlement was in the range of 2–4 mm and was virtually complete in the first day. Water was then added to the area on several occasions, but no further movement of the raft was detected.

Despite the low permeability, the pore pressures in fly ashes dissipate quickly. In triaxial tests[1] on 38 mm diameter, saturated specimens, the pore pressures dissipated in <10 min. This is consistent with the rapid settlements noticed above and indicates that long-term consolidation is not a problem, as settlement within the fly ash will occur during construction.

Shear strength

It was noticed very quickly that when fly ash was excavated from lagoons and stockpiles it could maintain a very steep, even vertical, face. The stockpile of fly ash excavated from lagoons at Carrington power station had side slopes typically of 45°. This highlighted the high shear strength possessed by fly ash. It was also noticed that strength could develop with time and was influenced by moisture content, with the strength falling significantly when the moisture content exceeded the optimum. A typical example is shown in Fig. 4.6. Data from Raymond[18] shows the effect of time and moisture content on the California bearing ratio (CBR) of fly ash from Bold power station.

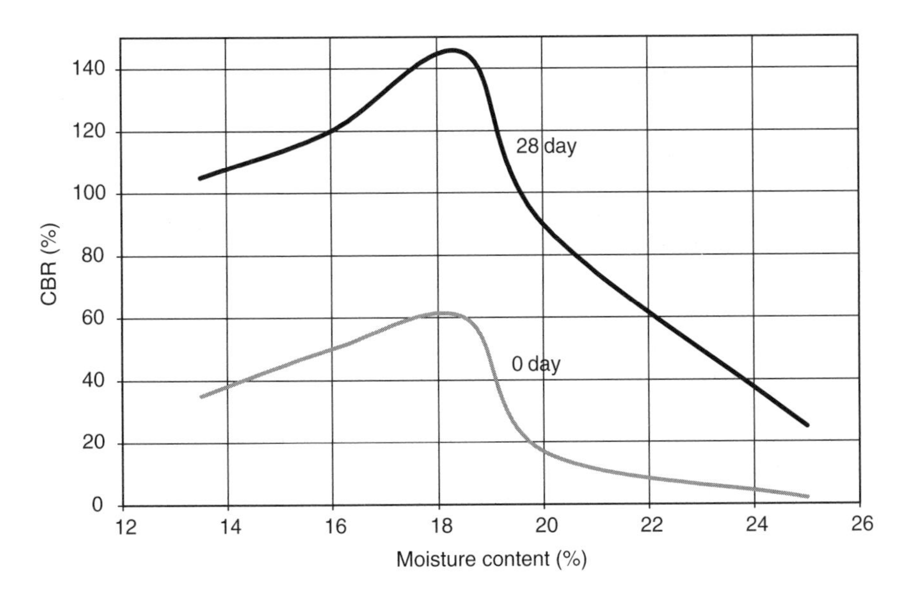

Fig. 4.6. Effect of time and moisture content on the California bearing ratio (CBR) value (Raymond[18])

Table 4.2. Variation in C_u and ϕ_u with time (Raymond and Smith[16])

Source	Elapsed time at test (days)					
	0		14		56	
	C_u (kPa)	ϕ_u (degrees)	C_u (kPa)	ϕ_u (degrees)	C_u (kPa)	ϕ_u (degrees)
Agecroft	45	33·6	207	34·6	207	37·0
Battersea	48	38·0	83	36·5	110	36·5
Bold	34	34·6	234	38·6	276	38·8
Dunston	41	33·6	55	34·2	69	34·7
Skelton Grange	28	34·3	110	40·0	179	40·0
Westwood	34	31·8	48	31·8	83	36·5

Early work examined strength by either CBR, unconfined compressive strength tests or undrained triaxial tests. Raymond[18] noted that it was very difficult to saturate the specimens when carrying out triaxial tests, recording very low *B*-values in tests on fly ash from a trial embankment in Bedford, UK.

Both Raymond[18] and Sutherland *et al.*[2] demonstrated that most fly ashes would gain strength with time but that this varied not only with source but also over time for a given source. The ratio of 28 to 1 day strengths ranged from 1·8 to 7·2 for Barony fly ash but from 0·9 to 2·1 for Kincardine. As discussed earlier, the strength gain is due to the chemistry of the fly ash, which in turn is dependent on the coal being burnt. The fly ashes with higher calcium contents show greater strength gain. Fox[19] confirmed the above, commenting that lagoon fly ash, where the water has been discharged from the lagoon, removing the water-soluble fraction, shows only moderate strength gain.

The strength gain is due to increases in both cohesion and friction angle. The increase in cohesion is due to the crystal growth and gel formation described earlier. The increase in friction angle is due to the rougher surface found on the fly ash particles as the self-hardening occurs. Undrained triaxial tests indicate that typically the cohesion will often more than double from 1 to 28 days, whereas the friction angle will show a rise of between 10% and 20%. There will be continuing strength gain beyond 28 days. Data obtained by Raymond and Smith[16] shown in Table 4.2 demonstrate the strength gains measured for different sources of fly ash. The difference in shear strength gain for fly ashes from Agecroft and Westwood is highlighted in Fig. 4.7. The two are similar at 0 days but there is a significant difference after 56 days, mainly due to the increased cohesion of the Agecroft fly ash.

As expected, the shear strength will reduce with density (Table 4.3). The cohesion at 90% relative compaction is typically 50–60% of that at 100% relative compaction, the friction angle typically falling by one-fifth.

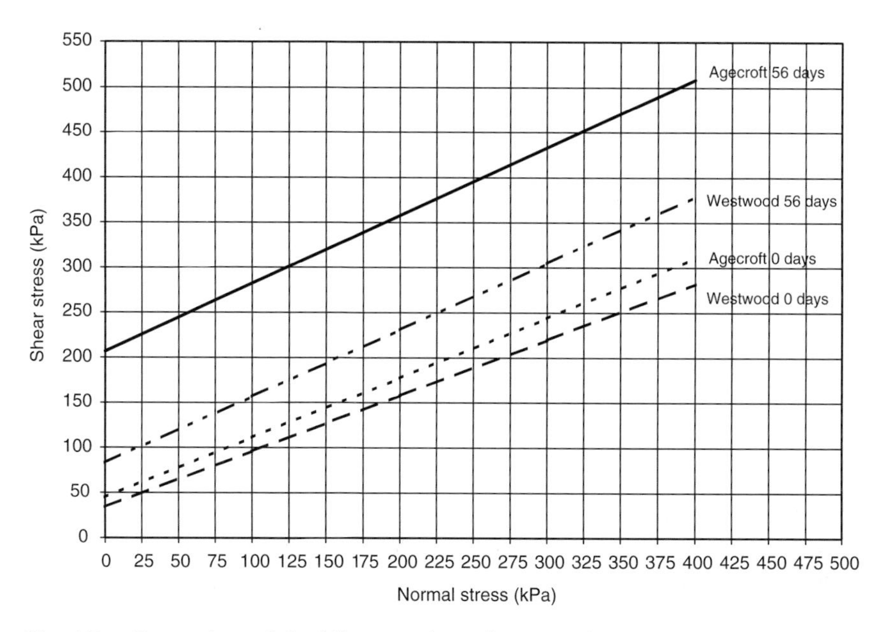

Fig. 4.7. Comparison of the failure envelopes for Agecroft and Westwood fly ashes at different ages (Raymond and Smith[16])

Table 4.3. Effect of compaction on shear strength (Sutherland et al.[2]*)*

Relative compaction (%)	Shear strength as a percentage of value at 100% relative compaction
85	60
90	75
95	90–95

The shear strength of fly ash falls when it becomes saturated. This is mainly due to a reduction in the cohesive element of the shear strength, with immediate tests on saturated samples sometimes indicating that there is no cohesion. This demonstrates that the immediate strength is largely due to suctions within the fly ash. There is less effect on the friction angle. This is shown in Fig. 4.8, which compares shear strength of a fly ash on a total stress basis, using samples tested at optimum moisture content without drainage, and strength on an effective stress basis from drained tests on saturated samples. Based on this, Coombs[20] suggested that a conservative estimate of effective shear strength can be made by assuming the friction angle to be the same as for total stress tests and to ignore the cohesion.

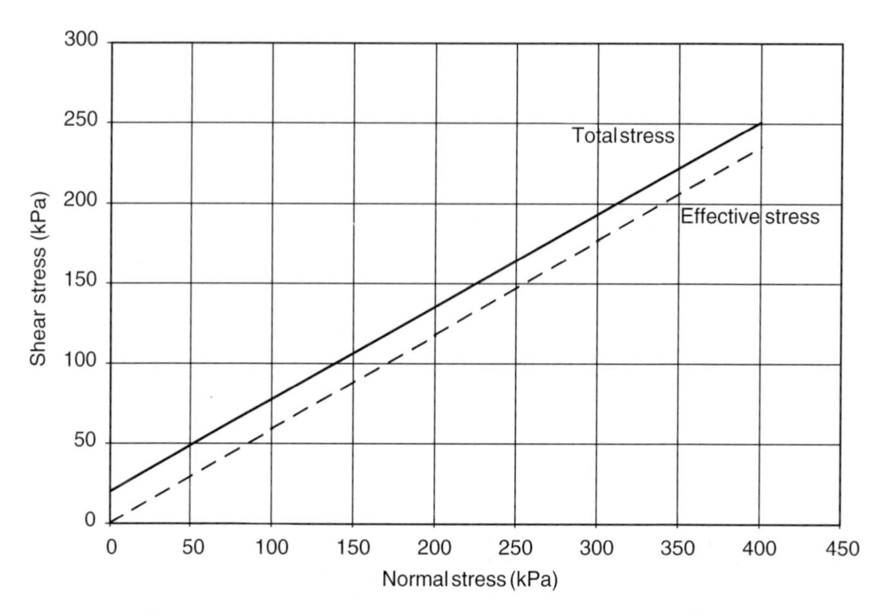

Fig. 4.8. Effective stress and total stress failure envelopes (Coombs[20])

Table 4.4. Peak shear strength parameters taken from direct shear tests on conditioned and stockpiled fly ash (Clarke and Coombs[21])

Age (days)	Conditioned fly ash				Stockpile fly ash			
	0	7	14	28	0	7	14	28
Unsoaked C' (kPa)	28	30	30	38	15	20	24	27
Unsoaked ϕ' (degrees)	43	41	43	44	44	44	41	43
Soaked C' (kPa)	2·4	8·1	8·7	2·8	1·7	2·7	5·1	7·8
Soaked ϕ' (degrees)	41	41	42	43	40	40	42	40

Although strength gain with time will be reduced, it often remains significant, as shown in (Table 4.4), based on work reported by Clarke and Coombs.[21]

As expected, the cohesion drops significantly when the fly ash becomes saturated, but there is some strength gain, although it is less than for the unsaturated material.

The effect of saturation of fly ash was examined and reported by Swain.[22] A trial embankment was constructed at Warrington New Town in an area with a high water table. This was drained and excavated, after which a small embankment was constructed in the area. The area was then allowed to flood. A series of samples was taken from the embankment to assess the

Fig. 4.9. Compacting fly ash as a fill material on the A52, UK

effect of saturation and these were compared with the performance of laboratory-prepared and -stored specimens. The results indicated that the laboratory specimens gained strength with time, mainly because of an increase in cohesion. The specimens taken from a site showed a reduction in strength as the material became saturated, although the undrained tests indicated that some cohesion was maintained. It was noted that some areas close to the edge of the embankment, where compaction was difficult, consolidated and after 151 days, the areas could safely take the weight of a man even though the fly ash was wet.

It is important to select the correct shear strength parameters when fly ash is to be used as a fill (Fig. 4.9). Saturated drained tests give results that may be too conservative in situations where the fly ash is unlikely to become saturated.

Reinforced earth

The first recorded use of fly ash in reinforced earth structures in the UK was on the Dewsbury ring road in 1988.[23] The poor ground in the area meant that alternatives were sought to conventional reinforced concrete retaining walls and the use of reinforced earth walls, up to 8 m high, with

Fig. 4.10. Fly ash as reinforced earth on the Connaught Crossing, UK

fly ash as the structural fill was considered the most suitable solution. The relatively low bulk density of fly ash was a major factor in the decision to use it. Before fly ash could be accepted it had to be subjected to an extensive testing regime to ensure that it could meet the Department of Transport's requirements. This involved building a trial wall, and the results showed that fly ash could perform as well as traditional fills (Fig. 4.10).

One of the main findings[24] was that although the redox potentials of fly ash were around 0·50 mV, the resistivity was typically in excess of 2000 Ω·cm, meaning that it could not be used with steel reinforcement. This is in agreement with work elsewhere[25] that indicated that the corrosion rate for steel embedded in fly ash was high. However, work by Braunton and Middleton[26] on the effects of fly ash on mild steel suggested that long-term corrosion rates are not linear, as is often supposed, but are parabolic with the depth of corrosion, as given below:

$$d = k \cdot \sqrt{t} \tag{4.1}$$

where d = depth of corrosion, k = constant and t = time.

The rate of corrosion will therefore fall with time, this being considered to be due to corrosion products inhibiting further attack from the surrounding material. Thus, short-term tests using linear corrosion rates would lead to an overestimate of the depth of corrosion. They

Table 4.5. Coefficient of interaction between fly ash and reinforcement (Jones[24])

Reinforcement	Coefficient of interaction
High-density polyethylene (Tensar SR2 geogrid)	0·81
Glass-reinforced polyester (Fibretain)	0·75

concluded that the fly ash used in the trial wall would have the required design life.

However, the findings were too late for the trial wall, so non-metallic reinforcement was examined; this included a high-density polyethylene and a glass-reinforced polyester. The coefficients of interaction between the reinforcement and the fly ash, measured using the method detailed by Sarsby[27] using a modified shear box, are shown in Table 4.5. The coefficient of interaction between fly ash and Paraweb has been found to be similar to that for Tensar. The success of the trial resulted in the construction of the walls at Dewsbury and also the inclusion of fly ash in the Department of Transport design code as an acceptable fill for reinforced earth.

Fly ash has been used successfully not only in reinforced earth walls but also in other projects. Where reinforcement has been used to increase the angle of embankments to reduce land take, polymer reinforcement has been used, although polyester has also been used in some structures. This reinforcing material is pH sensitive and generally values in excess of 10 are not acceptable. Because the pH of fly ash can be in excess of this, any such requirements need to be highlighted so that the fly ash supplier can ensure that suitable material is available. Lagoon fly ash would be useful in such circumstances because of its lower pH.

Miscellaneous uses

Fly ash has been used in some instances to improve the strength or handleability of soils. Very low moisture content fly ash can be added to wet soils or to improve the grading of single size sands. One example is the A27 at Avisford.[27] Here, fly ash was used to improve the moisture content of wet, clayey sands so that the material could be properly compacted.

Summary

Fly ash has been successfully used as both a general and a structural fill for over 40 years and has been demonstrated to be a valuable resource in construction, with the advantages of:

- low density
- high shear strength
- no consolidation.

The guidelines for successful use were established early in its history[28] and, provided these are followed, it will always prove to be a useful, forgiving material that even if mishandled can give good service.

Laying and compacting fly ash on-site as a fill material

The following are guidelines for laying and compacting fly ash for fill applications.

Recommended plant

- Spreading: Flat-tracked dozer (Drott or similar).
- Rolling: Towed or self-propelled vibrating roller, e.g. Bomag 90 or larger according to site.
- Small tools: Tarmac rakes have been found more useful than shovels for hand spreading.
- Heavy, dead-weight, self-propelled, smooth-wheeled rollers are not recommended.
- Pneumatic-tyred rollers have been found to be suitable, but there is little experience of their use in the UK.
- Once on site the fly ash must be spread and rolled as soon as possible to avoid loss of moisture and consequent dusting.

Recommended procedure

The following procedure is recommended for fly ash.

1. The fly ash should be spread in layers (recommended ~225 mm in the loose state) and be well 'tracked' with the spreading plant.
2. Every effort should be made to add sufficient water to the fly ash at the point of loading, but should it be too dry, it should be sprayed with water during tracking and before rolling. Stockpile fly ash will normally require the addition of further water after delivery, especially in windy/drying weather conditions.
3. If material is stockpiled the amount should be kept to a minimum and should be sprayed with water as required to prevent dust problems.
4. It has been found from experience that the moisture content of the fly ash can be roughly checked by visual inspection. Fly ash moulded in the hand should keep together in one mass when slight

pressure is exerted; when it is approximately at the correct moisture content, no moisture should be squeezed out. The moisture content can easily be measured using a 'Speedy' moisture meter. Microwave ovens have also been used, but care must be taken that the sample is not 'overdone', which may result in combustion of the carbon in the fly ash.

5. The fly ash should be spread in loose layers ~225 mm thick, compacting to 150 mm thick. On large sites, this is usually done with a bulldozer. Thicker layers are possible, but it is recommended that a trial is done to confirm the effectiveness of compaction (Table 4.6) for suitable types of plant. On confined sites where access is limited

Table 4.6. Suitable compaction plant

Type of compaction plant	Best location	Remarks
1. Allam Rampactor 2. Wacker rammer	Very confined areas, e.g. the narrow strip next to bridge abutments, retaining walls, underpasses	(a) The largest size shoe should be used (b) The moisture content and layer thicknesses must be correct to ensure adequate compaction
3. Tandem vibrating	Small and medium sized areas: (a) behind bridge abutments and retaining walls (b) structural fills to buildings, etc.	Bomag models smaller than those recommended are considered too light for fly ash Best results are obtained if the surface of fly ash is thoroughly 'tracked' by the spreading plant prior to compaction and the initial roller pass is without vibration. Usually eight roller passes are sufficient
4. Towed vibrating: Vibroll type T182 or similar 5. Self-propelled, pneumatic-tyred roller types: Albaret Autopactor Albaret – Unipactor Blaw Knox – Salcro	Large areas, e.g. embankments and other large open sites	The surface of the fly ash should be thoroughly 'tracked' by the spreading plant before self-propelled pneumatic tyred rollers can operate successfully

and a vibrating plate is used for compaction, thinner layers may be required.

6. The rolling should consist of no fewer than eight passes of the vibrating roller. The first two passes should be without vibration, the remaining passes should be with vibration on and the final pass should be in such a direction that the surface cracks are tightened up (this is usually a reverse pass, but depends on the slope). Sometimes a final pass with the vibration switched off will assist in closing up surface cracks.
7. Density testing can monitor the performance of the placing and compacting of fly ash. Because of the disturbance to the upper layer, density tests should be carried out in the penultimate layer. The core-cutter method has found to be reliable for this. Nuclear density meters have been used but the carbon content of the fly ash can influence the results. It is recommended that any density results be checked against oven-dried tests on occasion.
8. It is important to protect side slopes as soon as possible after completion of the fill operation to prevent scour in case of heavy rain. If the working area becomes saturated the water will not penetrate significantly and the fly ash will dry out rapidly if left. If the area needs to be worked before the fly ash is allowed to dry out then it can be bladed into a stockpile to dry and re-used later.

References

1. Yang Y, Clarke BG, Jones CJFP. A classification of pulverised fuel ash as an engineering fill. *Proceedings of the Conference 'Engineered Fills '93'*, Newcastle upon Tyne University, 1993.
2. Sutherland HB, Finlay TW, Cram IA. Engineering and related properties of pulverised fuel ash. *Journal of the Institution of Highway Engineers* 1968; June, 1–16.
3. Foreman R. The production of fly ash. *UKQAA Seminar*, Electricity Association, London, February 2000.
4. BS 1377. *Soils for civil engineering purposes, Part 3, Chemical and electrochemical tests*. BSI, London, 1990.
5. Sulfate and acid resistance of concrete in the ground. *Building Research Establishment Digest* 363, Watford, UK, July 1991 (NB. This document is being revised).
6. BS 1377. *Soils for civil engineering purposes, Part 4, Compaction related tests*. BSI, London, 1990.
7. Smith PH. Field trials on fly ash. *Contract Journal* 1962; September.
8. Meyers JF, Pichumani R, Kapples BS. *Fly ash as a construction material for highways*. US Department of Transportation, Washington, DC, 1976.
9. Cotton RD. Construction of embankments. *Proceedings of 'The use of fly ash in construction'*, Dundee University, 1992.

10. Margason G, Cross JE. *Settlement behind bridge abutments*. Ministry of Transport, RRL Report No. 48, 1966.
11. Barber EG, Jones GT, Knight PGK, Miles H. *Fly ash utilisation*. CEGB, London, 1972.
12. Oakham Ness Tanker Terminal: PFA for Local Bearing Fill. CEGB Technical Bulletin No. 11. CEGB, London, 1966.
13. *Pulverised fuel ash transport*. CEGB datasheet. CEGB, Sheerness.
14. Croney D, Jacobs JC. *The frost susceptibility of soils and road materials*. Ministry of Transport, RRL Report No. 90, Crowthorne, 1967.
15. Sutherland HB, Gaskin PN. *A laboratory investigation of the frost susceptibility characteristics of pulverised fuel ash*. Report No. 01038/4, University of Glasgow, 1967.
16. Raymond S, Smith PH. Shear strength, settlement and compaction characteristics of pulverised fuel ash. *Civil Engineering and Public Works Review* 1966; October.
17. Cabrera JG, Braim M, Rawcliffe J. The use of pulverised fuel ash for the construction of structural fills. *Proceedings of 'AshTech '84' 2nd International Conference on Ash Technology and Marketing*, London, 1984.
18. Raymond S. Pulverised fuel ash as an embankment material, *Proceedings of the Institution of Civil Engineers* 1961; **19**: 515–536.
19. Fox NH. Pulverised fuel ash as structural fill. *Proceedings of 'AshTech '84', 2nd International Conference on Ash Technology and Marketing*, London, 1984.
20. Coombs R. A comparison of properties of fresh and lagooned fly ash. Dissertation for Postgraduate Diploma in Geotechnical Engineering, Bolton Institute of Higher Education, 1987.
21. Clarke BG, Coombs R. Pulverised fuel ash as an engineering fill. *Proceedings of Bulk 'Inert' Waste: An opportunity for use*, Cabrera JG, Woolley GR, (eds), Leeds, 1995.
22. Swain A. Field study of the behaviour of pulverised fuel ash under partially saturated conditions. *Proceedings of the 1st International Ash Marketing and Technology Conference*, London, October 1978.
23. Jones CJFP, Cripwell JB, Bush DI. Reinforced earth trial structure for Dewsbury ring road. *Proceedings of the Institution of Civil Engineers, Part 1*, 1990: **88**.
24. Jones CJFP. The use of ash in reinforced earth. *Proceedings of 'AshTech '84', 2nd International Conference on Ash Technology and Marketing*, London, 1984.
25. Headon AC, Chan HT. Laboratory corrosion studies of metals in coal ash. *Proceedings of the 6th International Ash Symposium*, Reno, 1968.
26. Braunton PN, Middleton WR. Assessment of the corrosion of mild steel in reinforced earth structures back-filled with pulverised fuel ash, *Proceedings of 'AshTech '84', 2nd International Conference on Ash Technology and Marketing*, London, 1984.
27. Sarsby RW. The interaction between pulverised fuel ash and grid reinforcement. *Proceedings of 'AshTech '84', 2nd International Conference on Ash Technology and Marketing*, London, 1984.
28. CEGB Technical Bulletin No. 19 – A27 Trunk Road Improvement. CEGB, London, 1967.

Chapter 5

Use of fly ash for road construction, runways and similar projects

Introduction

Fly ash can be used in the construction of roads, runways and similar projects in a variety of ways. These can range from soil stabilisation and sub-base through to a constituent of the wearing surface. Fly ash as a constituent in concrete as a wearing surface was covered in Chapter 3. However, in the other applications the physical and pozzolanic properties of fly ash, as described in Chapter 1, are important. As previously described, a source of calcium hydroxide such as 'quicklime', hydrated lime or the by-product of the hydration of Portland cement provides the alkali for a pozzolanic reaction to occur. However, with many of these techniques, the initial strength of the system relies on the mechanical properties of the mixture. The ultimate strength and durability result from the pozzolanic reaction binding the various components together chemically or improving the bond formed, e.g. as when used with Portland cement.

The techniques described are kept as discrete methods. However, in general terms success depends on a few basic requirements, many of which apply to all material combinations:

- A mix design which properly caters for the particle size distribution of all the constituents. This is the only way to produce a dense and stable matrix.
- A mix design which is able to be compacted by the available plant. Some types of plant suit some applications and material combinations better.
- Plant and equipment which is capable of producing a fully integrated mixture of the cohesive fine powders involved, such as Portland cement, lime and fly ash. Often cement/lime contents are very low and effective and thorough mixing is imperative to guarantee the performance of the method.

- The mixes must have access to moisture and a source of calcium hydroxide. These requirements are vital in ensuring a pozzolanic reaction takes place.
- The application and techniques employed must be tolerant of the temperature/time periods needed for the pozzolanic reaction to occur.

These requirements are valid for a whole range of fly ash techniques, as follows.

General principles of stabilisation

Soil stabilisation is defined as the treatment of a material to improve its strength and other physical properties. While the properties of treated fly ash are adequate for many applications in civil engineering, the idea of stabilisation has figured prominently in its development. Many stabilisation techniques rely on reducing the water content of the *in situ* soil and increasing the strength and stability. The latter may be provided by a pozzolanic reaction between lime and a siliceous material, e.g. clay. However, for soils that are devoid of all suitable siliceous material, the addition of fly ash will provide the necessary pozzolana.

Development and application of stabilisation

Portland cement is probably the most widely used stabilising material. The resultant stabilised material, usually known as soil cement, has given good results in many parts of the world, especially in road and aircraft runway bases. A comprehensive cement technology has been developed as an offshoot of the parent science of soil mechanics.

Other additive stabilisers that have been relatively widely used are lime and bitumen. Quicklime is often used to stabilise soft, clayey soils. The reaction of quicklime and water produces hydrated lime and heat. This process helps with drying the soil and, when the treated material is compacted, forms a firm working platform for following construction. The lime both carbonates and reacts pozzolanically with the clay to give increased strength in the longer term. The choice of additive in any particular instance depends on the cost and availability of the additives in relation to the material to be stabilised.

Stabilisation processes

Soil stabilisation techniques have one thing in common in that the stabilised material is formed by intimately mixing a predetermined amount of additive with the soil and then compacting so far as possible under optimum conditions. In some parts of the world, the only materials available for road construction are the native soil and a limited choice of additives. These are special cases, however, and more typical are the conditions in the UK, where native materials and processes are available. Soil stabilisation using fly ash has to compete with other processes, materials and cements, for example, and must be cost effective.

Mix-in-place stabilisation

The mix-in-place method involves the mixing of the additive, soil and possibly water using a rotovator or specialised self-propelled pulveriser. Good mixing-in of the additive is possible with most soils. However, in the field the efficient mixing of additives with clay soils presents the greatest difficulty. Both Portland cement and quicklime have been used for mix-in-place stabilisation. Most of the work has been in areas of predominantly sandy soil. In such regions, the mix-in-place process will normally be the cheapest method of stabilising the soil.

Pre-mix stabilisation

The additive is mixed with the soil in separate mixers, in much the same way that concrete is mixed, so that much better control of mixing efficiency and quantities of water and additive is possible. Equipment similar to a concrete-mixing plant is required, e.g. weighing, batching and mixing systems are needed. Although it is possible to treat the existing soil in a road subgrade in this manner, selected soil from borrow pits is often used. Because of the extra excavation and haulage involved, this procedure is generally dearer than mixing in place, but may still be economical compared with other types of construction. A limiting factor in the pre-mix method tends to be the haulage distance of the borrowed soil. Since this is often relatively long in areas covered by clay soils, consideration began to be given in the 1950s to the use of fly ash as the imported material. The power stations producing fly ash are well distributed throughout the UK and they tend to be concentrated in those areas in which constructional activity is high. Hauls are thus not excessively long, and the material is low in cost.

Use of fly ash in stabilisation

Within stabilisation, fly ash can be considered to act in the following roles:

- As a stabilising agent: a material that can be used by itself to improve the physical properties of a soil or in conjunction with lime or cement to form a binder. The term 'soil' in this context includes imported material.

- As a binder as fly ash is a pozzolana: i.e. it will react with lime and water to form cementitious material. In the case of cement it reacts with the lime liberated during hydration but the product gains strength more slowly than the cement itself.
- As an aggregate: it has been shown that most fly ash can be successfully stabilised using economic amounts of lime or cement. In view of the lower gain in strength with lime and its greater sensitivity to low temperatures, cement is often preferred.

Stabilised fly ash can be used for

- road bases and sub-bases
- hard shoulders
- site roads
- footpath bases
- factory floors, hard standings, etc.

Although stabilised fly ash would not normally be considered a wearing surface it has been used successfully for coal-stocking areas in power stations.

Cement stabilised fly ash

General

Cement and lime are the most widely used soil-stabilising agents and most fly ash stabilisation contracts in the UK have been carried out with these as the binder. Fly ash, cement and lime are all readily available and of predictable performance. Therefore, fly ash can be considered as an aggregate in place of the existing soil in cases where this, owing to its chemical or physical properties, is unsuitable for direct stabilisation.

Design of cement fly ash or lime fly ash mixes

With cement fly ash (CFA) mixes, an increase in cement content gives an increase in strength, but since a pozzolanic material is being considered, the minimum cement content to give durability in the field should be used. Typically, 7% cement by weight can be used for sub-base and road-base applications, although mixes with up to twice this level have been used. For design purposes the unconfined compressive strength, or crushing strength, forms a useful indication and is used in the UK. It used to be generally accepted that the minimum crushing strength should be 1·8 MPa at 7 days, but many authorities now consider that it should be 2·8 MPa at this age. The upper limit of about 5·6 MPa should also be preserved, since at higher strengths the CFA tends to behave as a low-quality concrete in which cracks develop owing to lack of aggregate interlock. In good-quality CFA, the finer cracks that form do not impair the performance of the material.

With lime fly ash (LFA) mixes as little as 2–4% lime has proven to be effective and little improvement in compressive strength is found with higher lime contents. However, with such small quantities of binder effective mixing is very important. As indicated below for LFA mixtures, the rate of strength gain is significantly slower than when using Portland cement. For this reason LFA mixtures rely on the unbound strength for their early age performance characteristics.

Fly ash mixes have been improved by inclusion of a coarse aggregate, as in granular fly ash (GFA) below, and this is considered desirable if a supply of suitable material is available. Finer ash and granulated slag may also be used. The proportion in the mix is not critical as this will depend mainly on the grading of the coarse aggregate and should be determined by laboratory trials. The usual amount is ~25% of the fly ash by weight. The incorporation of a proportion of coarse aggregate will improve the mixing with cement and result in easier compaction. Furthermore, the mix also appears to be less affected by heavy rain if this occurs shortly after laying.

Influence of compaction

Compaction should be such as to give the maximum dry density for the compactive effort employed, which means that the moisture content must be controlled at the optimum or slightly above that required for the CFA mixture. Maximum dry density and optimum moisture are determined using standard compaction tests as described in Chapter 4. If mixtures of CFA or LFA are compacted drier than optimum using a vibratory roller there is a risk of overshearing the surface, leaving lenses of loose material. Unlike unbound fly ash, such material may not be properly incorporated when the next layer is placed. Reasonable correlation exists between laboratory optimum moisture content, the field value and the CFA. In the field, normally a minimum density of 95% of the compaction to refusal density should be specified.

Durability under frost action

Since fly ash is composed mainly of particles lying in the silt range, it is potentially frost susceptible, and many ashes suffer surface heave under freezing conditions. CFA with a cement content sufficient to give a 7 day compressive strength in the order of 2·8 MPa will not, however, undergo any significant volume change under frost action. LFA mixtures, owing to their slower strength gain, may be adversely affected by frost. If they are protected by 450 mm of surfacing material this will offer sufficient protection. However, some assessment of frost heave properties may be required where less cover is possible.

Soluble sulfate content

Most ashes, unless lagooned, have soluble sulfate contents that are high compared with soils. The acid-soluble extract from 13 ashes gave a value of between 0·49% and 1·42% SO_3, with the majority present as gypsum. These can be taken as typical values. Despite this, the set of the cement and gain of strength time are unaffected for mixes cured in the partially saturated condition.

Field procedure and plant

The field procedure chosen will depend on the plant available, and the possible methods are:

- mix-in-place
- multi-pass plant
- single-pass plant
- pre-mix.

Mix-in-place methods: general

Mix-in-place methods will give a finished thickness of up to 200 mm, although 150 mm is more usually specified. After preliminary excavation has been carried out, the fly ash is spread at the required thickness by a blade grader over the area to be stabilised. In the normal delivered condition, 200 mm thickness of loose material will reduce to 150 mm after compaction. Careful control is essential at this stage to give a regular surface and accurate levels on the finished layer. The cement or lime is then added either by mechanical spreader (which can be adjusted to give the required distribution) or by the following manual method. The surface of the area is pegged out into rectangles of such area to require 50 kg of cement to give the correct mix. A bag of cement is then placed in each rectangle. The bags are then broken and the cement is raked evenly over the area.

Multi-pass methods

After cement or lime spreading, an initial mixing is given to prevent blowing and any necessary adjustment to the moisture content is made. Water is supplied by tanker, preferably with an offset spray bar so that the tyres do not compact the mixed material. Quicklime may be used on wet sites. The reaction between quicklime and water reduces the moisture content of the site. Where added, the amount of water applied to the fly ash is controlled by the pressure at the spray bar with the tanker running at a constant speed. If a large increase in moisture content is required, the process could be repeated between passes of the mixing equipment. The tines of the machine break up any lumps in the fly ash and mix in the cement or lime to the required degree by a series of passes over the layer. Power is supplied through a coupling from a take-off from the towing tractor.

Single-pass methods

Single-pass plant differs from multi-pass mixing by employing a specialised rotary tiller with a high speed of rotation and a very low forward speed. A range of proprietary equipment is available, specifically designed to ensure efficient mixing without creating dust problems.

Compaction

Here, independent compaction plant is needed; the first pass is preferably carried out by tracked equipment as this gives a very good surface on which following plant can operate. In general, good compaction and a good surface finish can be achieved by using plant described in Chapter 4 and this chapter. Choice will depend on the scale of the job and the number of stabilising machines used.

Pre-mix methods

Although very efficient and economical stabilisation can be carried out with plant of the type described above, pre-mix methods cannot be ignored. The fly ash has to be brought to the site, and pre-treatment in a stationary mixer enables good control to be kept over the moisture and cement or lime contents, and the efficiency of mixing. The mixer should be specially designed to prevent the formation of pellets of unmixed fly ash. Either high-speed double-shaft continuous mixers or high-speed pan mixers may be used. The configuration and mixer blade settings can have a significant effect on the quality of the material. Balling of the mixture can be caused by badly adjusted or worn blades and regular maintenance is essential.

A works trial should be carried out in each case as the machine has to mix two fine and uniformly graded powders with water. The CFA is spread to correct levels and compacted as in the mix-in-place method. Hand laying may be used for small jobs but for large schemes machine laying, using e.g. a paver machine, will be more economical and will give an improved profile.

Multilayer construction

Pre-mix construction has marked advantages where thick layers of stabilised material are required, as for example in a construction of bunds, bridge seating and road bases of thickness >150–200 mm. Actual work *in situ* is restricted to spreading and compaction, the efficiency of the latter governing the thickness of the layer that can be used. Although multilayer work is possible with the other procedures, there is a risk of debonding occurring at the interfaces of the layers.

Influence of weather

The usual precautions against the effects of adverse weather should be taken as in other forms of construction. For example, adequate cover must be provided in frosty weather. In wet conditions the laying of polyethylene sheeting on the subgrade, despite adding to the cost, helps to protect the soil and prevents wetting-up of the underside of the CFA by absorption of soil moisture. In mix-in-place construction, the water supply must be adjusted to make allowance for any increase in ash moisture due to rainfall. Pre-mix construction suffers least from rainy conditions so long as the compaction process follows quickly after the spreading. It may be advantageous to use rapid-hardening cements to improve early strength properties. The gain in strength gives added safeguard against the effects of adverse weather after compaction, at small additional cost.

Field control

With soil stabilisation in general, good site provision and control testing are essential for CFA or LFA construction. The controls required vary with the nature of the job.

Curing of fly ash/cement or lime

It is important that the compacted CFA or LFA is maintained at the 'as-laid' moisture content for as long as possible to allow the development of strength. Sisal or hessian covers can be used, but the best method of preventing a rapid drying layer is to spray the green surface with bitumen sealer. This provides not only a proof membrane, but also a non-strip layer on to which the subsequent surfacing can be readily laid.

Road bases and sub-bases produced with fly ash bound mixtures

The following text describes the techniques that are specific to road base and sub-base construction. They have been developed in the UK based on experience and usage in France. Many French roads, including the heavily trafficked AutoRoutes, use fly ash as a cementitious binder/aggregate in their construction. In recent years in Europe these systems have become known as fly ash bound mixtures (FABMs). FABM is a construction material for road and airfield pavements. It is a mixture of fly ash and one or more other components, the performance of which relies on the pozzolanic properties of the fly ash. As the pozzolanic properties require the presence of an alkali activator, normally lime, a wide range of fly ashes may be used successfully.

Types and composition of fly ash suitable for fly ash bound mixtures

Dry, conditioned or lagoon fly ash can be used for FABM. The fly ash need not be fresh as the pozzolanic properties depend on the presence of an alkali to initiate the reaction. Old, stockpile fly ash may be used and found perfectly acceptable. Indeed, it maybe advantageous in some respects to use stockpile ash in preference to fresh material. Significant proportions of UK ash stocks are suitable for making FABM. The following are the basic requirements from the fly ash:

- *Particle size:* Carried out in accordance with BS EN 451-2, shall conform to those in Table 5.1.
- *Chemical composition:* Expressed as a percentage by mass of the dry product, which is obtained by drying a laboratory sample in a well-ventilated oven at 105 ± 5°C to constant weight, and cooled in a dry atmosphere.
- *Loss on ignition (LOI):* The LOI, measured in accordance with BS EN 196-2, but using an ignition time of 1 h, or other equivalent method, shall not exceed 8% by mass. If, proportion-wise, the fly ash is the main component in the FABM then the LOI shall not exceed 10%.*

Table 5.1. Particle size limitations for fly ash

Sieve size (μm)	% by mass passing
90	⩾70
45	⩾40

*The purpose of this requirement is to limit the residue of unburned carbon in fly ash. It is sufficient, therefore, to show through direct measurement of unburnt carbon residue, that it is less than the value specified above.

- *Sulfate content:* The sulfate content, expressed as total SO_3, shall not exceed 4% by mass when measured in accordance with BS EN 196-2.
- *Free calcium oxide content:* The free calcium oxide content, measured in accordance with BS EN 451-1, shall not exceed 1% by mass. If this requirement is not met, soundness shall be measured in accordance with BS EN 196-3, and the expansion shall not exceed 10 mm with a 50 : 50 blend of fly ash and cement.
- *Water content:* Dry fly ash shall contain not >1% mass of water. Fly ash can be stored, used and supplied in either a wet or dry condition.

As with concrete, the binders can be either blended on site or produced in a factory and added as a hydraulic road binder.

Types of fly ash bound mixture

There are many types of FABM. They all rely on the pozzolanic reaction resulting from the combination of fly ash and the added lime or the by-product lime created when Portland cement hydrates. Examples of FABMs are shown in Table 5.2. This table has been extracted from a draft European standard, prEN13285 Part 3 'Unbound and hydraulically bound mixtures – fly ash bound mixtures', which is due to be published in 2003. It should be noted that the list of FABMs shown in the table is not intended to be exhaustive but illustrative of the current use of FABMs in Europe.

FABM, which is based on the addition of quicklime or hydrated limes, reacts slowly and the reaction rate is temperature dependent. This is advantageous for many applications. The initial strength in the layer of FABM is due to internal cohesion and friction rather than chemical bonding, as with cement bound materials. This allows freshly laid FABM to be trafficked without detriment to the long-term strength and stability of the material. The trafficking of the material also aids compaction. Granular fly ash (GFA) is particularly suited to immediate trafficking after compaction. Sand fly ash (SFA) mixtures can also be used subject to a suitable bearing index being achieved. For use in colder weather conditions and cooler climates lime-only based FABM may prove problematical and Portland cement can be used to increase the rate of hardening in these circumstances.

Manufacturing fly ash bound mixtures

With respect to the quality of the finished product, FABMs are preferably produced in central batching plants using pug-mill type continuous mixers (see Fig. 5.12). Other stationary mixers and the mix-in-place method of construction can be employed in certain situations. Owing to the cohesive nature of fly ash-based mixtures and the low water contents and workabilities involved, forced action mixers are preferred to ensure

Table 5.2. *Examples of fly ash bound mixtures (FABMs) for road and airfield pavements*

Type of FABM	Abbreviation	Typical proportions as a percentage of dry mass (%)								
		Conditioned fly ash	Lime (CaO)	PC	Graded crushed coarse material	Sand	Soil/ earth	Other material	Typical water content (%)	Normal age of performance testing (days)*
Lime fly ash	LFA	93–97	3–7						15–25	90
Lime gypsum fly ash		91	4					5% gypsum	15–25	90
Cement fly ash	CFA	90–95		5–10						28
Lime fly ash granular material (two options)	GFA	8·5–13	1·5–3		50–55	30–40			6–8	90
			$1^{\dagger}$		50–55	40–45		4–6% dry fly ash	6–8	90
Cement fly ash granular material		3–6		1–3	50–55	40–45			6–8	28
Slag fly ash granular material		5–7	0–2		50–55	30–40		5–7% GBS	6–8	90
Lime fly ash sand	SFA	9–12	2–4			84–89			~10	90
Cement fly ash sand		6–8		2–4		88–92			~10	28
Lime fly ash earth (soil)	EFA		$1–2^{\dagger}$				90–93	6–8% dry fly ash	Depends on soil	90
Cement fly ash earth		3–6		2–4			91–94		Depends on soil	28

*Earlier age testing is permissible subject to data and experience.
†Lime is usually preblended with fly ash.
GBS: granulated blastfurnace slag.

adequate dispersal of the relatively low lime or cement additions being used. For mix-in-place applications, the use of farming equipment has been superseded in recent years by specially designed machines. These reduce the problems associated with *in situ* mixing of fly ash, lime and cement, and the dust that can be created.

Laying

Placement and compaction are by conventional plant such as drot, grader, paver and vibrating roller. Pneumatic-tyred rollers are usually specified for finishing purposes and for some FABMs, as the only means of compaction. Immediately after compaction, FABM shall be prevented from drying out by the application of an alkaline bitumen emulsion or the repeated light-spray application of water. The slow rate of hardening of FABM ensures good workability and some capacity for self-healing. The mechanical interlock between the particles (with granular materials) and the good cohesion in LFA allow for immediate trafficability.

Pavement terminology

A road and airfield pavement construction consists of a multilayer system (Table 5.3).

Road design using fly ash bound mixtures

The conventional road design in the UK is a bituminous surfacing and road base over an unbound granular type I sub-base. Table 5.4 compares a traditional UK flexible design with FABM equivalent designs.

The capping layer thickness depends on the strength of the subgrade and can vary from nothing to 600 mm of material consisting of a wide range of materials. Furnace bottom ash has been used successfully as a capping layer in many European countries. However, stabilised fly ash can be used as a capping layer.[1] More recently, recycled road planings have been used for capping.

Applications of fly ash bound mixtures

FABMs may be used for capping layers, sub-bases and road bases of all classes of road and airfield pavements and footways. Fly ash is a pozzolanic material, which in the presence of lime [CaO, quicklime, or $Ca(OH)_2$, hydrated lime] hardens under water. Compared with ordinary Portland cement, the rate of hardening of the fly ash/lime combination is much more protracted, which has advantages in pavement construction.

Table 5.3. Multiple layers used in road construction

Layer		Description
Bituminous or cementitious surfacing	Wearing course	This is the layer used to provide the all-weather, wearing properties, skid resistance and texture, and a smooth-riding road surface. With bituminous systems, these materials are the most expensive in the road and very thin, e.g. typically 25–30 mm If a cementitious-based system is used the wearing course will normally be combined with the base-course layer to form a rigid, thick slab which is textured
	Base course	This layer is a level regulating layer and provides thermal insulation to following layers
Road base		The main long-term structural layer in the road. This could be either bituminous (HRBM), a cementitious lean concrete or a FABM such as GFA
Sub-base layer		This layer provides a working platform for contractors who need to work in all weather conditions. This is designed for a long life, typically 40 years. It may act as a drainage layer and provide a frost break. Type I crushed rock or FABM (GFA) can be used, although neither of these will act as a drainage layer
Capping layer		Similar to sub-base layer and only used when the subgrade is poor, i.e. crushed rock or FABM. Suitable FABMs would be CFA, LFA, EFA or SFA. May be a wide range of materials including granular, which can be asphalt planings and stabilised cohesive materials such as FABM. Open textured materials will act as a drainage layer; however, FABM is not an open textured material
Subgrade		The soil on which the road is formed

HRBM: hot-rolled bituminous material; FABM: fly ash bound mixture. For other abbrevia-

Table 5.4. Illustration of fly ash bound mixture (FABM) pavements and their design

UK traditional flexible design	FABM sub-base option	FABM sub-base and road-base option
30 mm SMA wearing course over	30 mm SMA wearing course over	30 mm SMA wearing course over
70 mm DBM base course over	70 mm DBM base course over	70 mm DBM base course over
200 mm DBM road base over	125 mm DBM road base over	175 mm GFA road base* over
150 mm type I sub-base	225 mm GFA sub-base	175 mm GFA sub-base*
450 mm in total	450 mm in total	450 mm in total

*These should be laid as two separate layers, compacting each layer separately.
SMA: stone mastic asphalt; DBM: dense bitumen macadam; GFA: granular fly ash mixture (granular material treated with fly ash/fly ash and lime).

- In the short term, FABMs have extended handling times and thus the flexibility in the construction process of unbound granular pavement materials, e.g. type I sub-base.
- In the long term, FABMs develop significant stiffness and strength, giving them the performance and durability of bituminous and cement-bound materials.

Where quicker hardening is required, e.g. in cold weather working, partial or complete replacement of lime with cement or the addition of gypsum or other suitable material can be performed.

Granular fly ash mixtures

Granular fly ash (GFA) is a mixture of crushed graded coarse material, sand, fly ash, lime or cement, possibly slag and water, where the fly ash and lime combination performs as a binder. GFA can be used for sub-bases and road bases of all classes of road and airfield pavements and footways. The GFA is laid on a subgrade, capping or sub-base material with a soaked laboratory California bearing ratio (CBR) of at least 15%.

Characteristics, performance and durability of granular fly ash

GFA is a cementitious material that changes from an unbound crushed stone material into a bound paving material, the rate of reaction being strongly

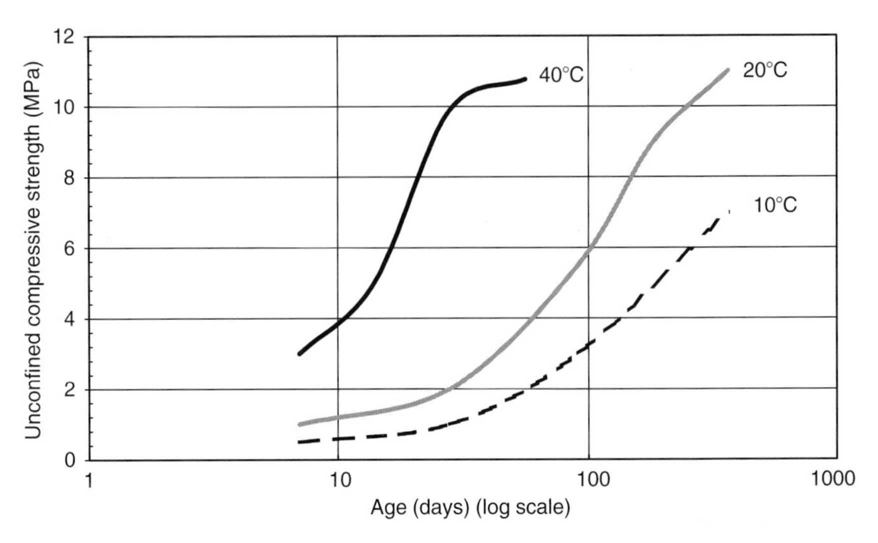

Fig. 5.1. Effect of temperature on unconfined compressive strength for a typical granular fly ash

dependent on temperature (Fig. 5.1). This has advantages in road construction under bituminous or pavement quality concrete surfacing. In the short term, GFA has a handling time of many hours with similar flexibility to unbound granular paving materials, e.g. type I. In the long term, and depending on the aggregate, GFA develops significant elastic stiffness (ranging between 10 and 30 GPa) and tensile strength (between 1 and 3 MPa), which results in a pavement material with the performance and durability of bituminous and cement-bound materials. The slow reaction rate means that there is extended time to work the material and the capacity to self-heal.

In the fresh condition on normal sites, the correct grading framework for the GFA produces a tight, closed finish that can withstand light rain. However, on hilly sites, the fines in the fresh GFA can be removed by running water and measures should be taken to prevent this occurring by the provision of adequate drainage systems. The use of crushed material and the correct grading framework is responsible for the immediate trafficability of GFA. Thus, the stability of GFA over the short and medium term favours the development of the strength and stiffness in the longer term. The ultimate structural characteristics may not be achieved until 2–3 years after laying. The slow reaction rate generally limits construction to the period April to October inclusive (in the UK) to enable frost resistance to be achieved before the first frosts. Outside this period, the setting time can be reduced by the partial or complete replacement of the lime with cement or the addition of an accelerator such as gypsum. Overlaying the material will give some protection from frost.

Structural design of granular fly ash pavements for roads

Basis of the design

These designs cater for traffic in excess of 80 million standard axles (msa), based on French design practice. These documents include designs for GFA up to the equivalent of 320 msa, the maximum catered for on French roads:

- 1977 Catalogue des structures types de chaussees neuves, DRCR, SETRA, LCPC
- 1998 Catalogue des structures types de chaussees neuves, DR, SETRA, LCPC.

When interpreting these documents to formulate the designs, account has to be taken of French practices such as:

- 13 tonne maximum axle load
- the different approach to capping design
- the differing pavement life/maintenance strategies.

Regarding axle loads, the designs offered below relate to a maximum of 11.5 tonnes. Using analytical design, this has the effect of reducing the French design thicknesses by 30–40 mm.

For capping, the French have four strength categories, PF_1, PF_2, PF_3 and PF_4. PF is the abbreviation for 'platforme'. The categories relate to formation surface stiffnesses of 20, 50, 120 and 200 MPa, respectively. For the purpose of this document, the top of the standard UK foundation of type I granular sub-base on capping has been assumed to lie between categories PF_2 and PF_3. The top of the standard UK capping is assumed to lie between categories PF_1 and PF_2.

The French pavement life/maintenance strategy in the above references includes constructions based on stage construction at 7–9 year intervals. In addition, they provide constructions that have structural design lives up to either 20 or 30 years, depending on the class of road. The designs employed are based on the equivalent of the '20 year-to-critical' design life normally used in the UK. They have been checked against UK practice for similar pavement performance and by the semi-analytical approach described by Williams.[2] Williams, in turn, based his approach on an empirical formula developed by Lister and Jones[3] (TRL).

Design for the maintenance and strengthening of new roads, including reconstruction and overlay

Table 5.5 has been formulated to satisfy both new build and maintenance scenarios. All that is required to use the table is knowledge of the design

Table 5.5. Design recommendations for granular fly ash (GFA) road construction

Traffic (msa)	Combined thin-wearing course and bituminous layer thickness (mm)	GFA thickness (mm) as a function of the surface stiffness or CBR of the underlying supporting layer/formation (MPa)					
		200 MPa	70 MPa	40 MPa	30 MPa	20 MPa	10 MPa
		Stabilised formation using a suitable depth of OPC, lime + OPC, lime + GGBS or lime + fly ash treatment	Equivalent to top of UK standard foundation of 225 mm type I sub-base on subgrade CBR of 5% or 150 mm type I granular sub-base on capping	Equivalent to top of standard UK capping	Equivalent to a CBR of ~5%	Equivalent to a CBR of ~3%	Equivalent to a CBR of ~2%
>80	170	230	260	290	330	380	430
30–80	170						
12–30	130	200	230	260	300	350	400
6–12	100						
2–6	100						
0·5–2	80	180	210	240	280	330	380
<0·5	60						

1. The design recommendations assume a GFA meeting the specification requirements and T_3 tensile strength class (or equivalent compressive strength class) of UKQAA data sheet 6.7.
2. In the case of GFA of T_2 tensile strength class, the GFA thickness shall be increased by 30 mm. In the case of class T_4, the GFA thickness can be decreased by 30 mm, except for a surface stiffness of 200 MPa, where the GFA thickness shall be unchanged.
3. In the case of frost susceptible subgrade materials or capping, the depth of overlying non-frost-susceptible construction shall satisfy local requirements.
4. Do not use the table to verify a correlation between surface stiffness and California bearing ratio (CBR). The table has been formulated to account for both the short- and long-term situation for the foundation under the GFA.
5. For traffic between 2 and 12 msa, and depending on individual circumstances, a reduction in the total bituminous cover is permitted provided the resulting bituminous cover is not less than 80 mm and the reduction is offset by an equal increase in the thickness of the GFA.
6. For traffic in excess of 12 msa, and depending on individual circumstances, a reduction in the total bituminous cover is permitted provided the resulting bituminous cover is not less than 100 mm and the reduction is offset by an equal increase in the thickness of the GFA.

OPC: ordinary Portland cement; GGBS: ground granulated blastfurnace slag.

traffic in msa and the support condition of the supporting layer/formation. Table 5.5 must be used in conjunction with the notes below the table. Under certain situations, such as patch and overlay jobs, it is possible to match the bituminous cover to the GFA given in the table with local standard overlay practice or requirements (see notes 5 and 6).

Lime fly ash and cement fly ash mixtures

LFA is simply fly ash with added lime either as CaO (quicklime) or as $Ca(OH)_2$ (hydrated lime). It is a slow setting and hardening mixture with self-healing properties. Table 5.6 shows the results from various LFA laboratory trial mixes that had differing lime contents with CFA mixes of similar composition for comparison purposes. The results show the advantage of Portland cement over lime at 7 days but illustrate the superiority of lime at 91 days. They suggest that 5% CaO is equivalent to 8% Portland cement. The soaked strengths for LFA are about 80% of the unsoaked strengths.

The cube strength requirement for cement bound sub-base (CBM I) for a flexible pavement is 4·5 MPa at 7 days. Projection to 91 days yields an equivalent strength requirement of ~7 MPa. LFA with 5% CaO satisfies this 91 day projection. The typical UK requirement for capping is a soaked CBR of 15%. The above results indicate that since the LFA mixture with 2·5% CaO is almost of CBM I quality and soaked strengths are good, it

Table 5.6. Strengths of various lime/fly ash (LFA) mixtures in comparison with Portland cement

Sealed specimens	LFA with 2·5% CaO	LFA with 5% CaO	CFA with 7% PC	CFA with 9% PC
7 days	1·5	1·8	3·0	5·0
28 days	–	–	4·0	8·0
35 days	4·0	4·0	–	–
28 days + 7 days in water	3·3	3·3	–	–
91 days	5·0	7·3	6·0	9·0

Standard Proctor optimum moisture content (OMC) for mixtures ~21%. Typical specimen wet density ~1600 kg/m^3.
Mixture percentages are based on dry weight. Thus, 2·5% CaO ~33 kg/m^3.
Strength results are for 1 : 1 cylinders and can be considered equivalent to cubes.
Specimens were cured at 20°C and sealed to prevent evaporation.
The results at 28 + 7 days designate 28 days' curing by sealing followed by 7 days in water.
LFA: lime fly ash; CFA: cement fly ash.

Fig. 5.2. Grading lime/fly ash to level

should satisfy capping strength requirements. The above results are typical for LFA mixtures. However, there will be some variation in strength for fly ash and lime from different sources. It should be noted that below 5°C, the reaction between lime and fly ash virtually ceases. This is generally not a problem with capping, but LFA sub-base work should be limited to warmer weather conditions, e.g. ambient temperatures of 10°C and more, unless the roadbase is laid and surfaced before the first frosts. Soft burnt fine-grade quicklime or hydrated lime should be used for LFA as hard burnt lime has caused problems with 'pop-outs' and expansion. This appears to be due to the hard burnt material reacting more slowly, resulting in gradual expansion of the compacted material as it fully hydrates. LFA is best produced in pug-mill mixers, laid 'high' and trimmed by 'tracked' blades (Fig. 5.2), and compacted by a pneumatic-tyred roller (Fig. 5.3). Alternatively, *in situ* stabilisation can be used. In general, LFA at optimum moisture content will support traffic immediately, although surface disturbance may occur, but this can be rectified with wetting, shaping and rolling 3 days and longer after laying. For best results, LFA should be overlain within 4 h by the next layer. If this is not possible, LFA should be sealed or kept moist to prevent drying out.

CFA can also be used for lightly trafficked roads in the same manner as LFA.

Fig. 5.3. Compacting lime/fly ash with a rubber-tyred roller

Pulverised fuel ash/fly ash in cement-bound material: CBM III, IV and V

Cement bound materials (CBM) III, IV and V are categories of Portland cement-bound materials that are specified in the UK Department of Transport's (DOT) *Specification for highway works* (SHW) for use in trunk road and motorway pavements. In accordance with the DOT's *Design manual for roads and bridges*, Volume 7:

- CBM III is the specified sub-base for concrete pavements across the full traffic range.
- CBM III, IV and V are permitted roadbases for traffic up to 80 million standard axles.

CBM III, IV and V are also specified by the relevant authorities for airfield and port pavements.

The use of fly ash in cement-bound material

The structural and cost benefits from fly ash use in CBM have long been recognised and employed. These benefits relate to the more progressive strength development that fly ash gives CBM. The strength gain beyond 7 days is usually much higher than with straight OPC mixtures. The Notes for Guidance on the SHW recognise this difference and suggest that when

Fig. 5.4. Mixing fly ash modified CBM IV (courtesy of Fitzpatrick Contractors Ltd)

fly ash is used in CBM, cube strength compliance should be carried out at 28 days rather than the 7 days for the usual Portland cement (CEM I) mixtures (Fig. 5.4). This is provided the contractor shows from trial mixes that the 28 day strength of the fly ash modified CBM compares with that of the CEM I mixture which meets the SHW requirements at 7 days. The Notes for Guidance also provide construction advice. The Notes for Guidance also provide construction advice. Figures 5.4 and 5.5 show fly ash modified CBM IV being used.

Strength compliance of CBM III, IV and V with fly ash

Table 5.7 shows results from the 1998 work using different sources of fly ash, Portland cements (CEM I) and aggregates.

The strength developments shown for CBM III, IV and V (CEM I) illustrate the theoretical minimum cases. The real results shown for the fly ash-modified CBM illustrate that it easily realises CBM III, IV and V PC strengths at 28 days.

Airport paving case study

The East Midlands Airport Project,* in the UK, required that an area some 15.6 ha (39 acres) be constructed for the use of various types of aircraft,

*The authors would like to thank the John E Ferguson of Fitzpatrick Contractors Ltd of Hertfordshire for their co-operation in preparing this case study.

Table 5.7 – Compressive strength results for cement bound materials (CBMs)

Mix designation	Proportions (% by dry weight)			Cube strength (MPa) at various ages (days)				
	CEM I	Fly ash	Aggregate	3	7	28	56	91
CBM III (CEM I): theoretical	5	–	95	–	⩾10	~12	~13	~14
CBM III (CEM I/fly ash): actual	3	12	85	5	6·5	15	19	22
CBM IV (CEM I): theoretical	6	–	94	–	⩾15	~18	~20	~21
CBM V (CEM I): theoretical	7	–	93	–	⩾20	~24	~26	~28
CBM IV/V (CEM I/fly actual	3·5	7·5	89	–	16	24	–	–

Fig. 5.5. Paving fly ash modified CBM IV roadbase (courtesy of Fitzpatrick Contractors Ltd)

ranging in size up to and including 747 jumbo jets. This involved the construction of a large turning apron, incorporating taxiways, and some 16 aircraft stands adjacent to the express cargo building, which was constructed under a separate contract. The client for the Project was East Midlands Airport Ltd, with Scott Wilson Kirkpatrick & Co. Ltd as their acting consulting engineers. The contract was awarded to Fitzpatrick Contractors Ltd, a company with many years' experience in similar projects world-wide. Burks Green Ltd was engaged as designers for the project. Construction commenced in October 1998. The paving element of the works started in April 1999 with a completion date of October 1999.

The existing subgrade was mainly a firm clay material and it was decided to lay 225 mm of dry-lean concrete (DLC) directly on to this formation.

Following extensive trials, the final mix design for the dry-lean incorporated 30% fly ash as part of the total cementitious content; 20–5 mm limestone coarse aggregate, 3 mm down limestone dust and medium grade concreting sand were used. The binders used were 91 kg/m^3 of Portland cement and 30 kg/m^3 of fly ash to BS EN 450 as supplied from Ratcliffe-on-Soar power station near Nottingham, UK. During the mix trials, it was planned to use a DLC with a high fly ash content. Compaction of this mix, owing to the high fines content contributed to by the fine aggregate, proved problematical and the option was shelved for further investigation later.

Fig. 5.6. Laying lean mix concrete

The density requirements were 95% of cube refusal density and the strength was in excess of 15 MPa at 7 days. The mix with 30% fly ash easily achieved these parameters.

Placing and compacting the dry-lean Concrete

The DLC was site batched using two mixing plants. The majority of the DLC was produced by an Erie Strayer 9 m^3 tilting drum mixer, backed up by a 2.5 cm Elba ESM 110 plant. The mixed DLC was transported to the point of deposition in Maxon Agitors, each having a capacity of 9 m^3. These trucks, together with all of the plant used on the project, are parts of an extensive fleet of paving plant owned and operated by Fitzpatrick. It was not necessary to use the agitator paddles in the truck bodies with the DLC owing to the free-flowing nature of the fly ash mix.

The DLC was spread on to the formation using an ABG Titan 423 tracked paver (Figs 5.6–5.8). The vibrating Duo-Tamp on this paver achieved 92–93% of the required compaction. A Bomag 135 tandem roller was used to complete the compaction and close the surface. Curing of the DLC was by conventional bituminous spray.

Pavement-quality mix design

The pavement-quality concrete (PQC) mix had to meet stringent flexural strength requirements. The minimum flexural requirement was 4·5 MPa at 28 days. In order to satisfy this requirement a strength of 5·1 MPa at 28 days

Fig. 5.7. Specialised tipping concrete trucks

Fig. 5.8. Spreading the concrete

was required *in situ*. Following extensive site mix trials it was concluded that a strength of 6·3 MPa at 28 days in laboratory-cured beams would satisfy the desired criteria. The average strength achieved for the contract was 6·9 MPa at 28 days and 8·7 MPa at 56 days. Compressive strength was closely monitored using test cubes throughout the works. Testing was carried out in accordance with the contract testing plan. Cubes for testing at 3, 7, 14, 28, 56 and 91 days were made at regular intervals. Beams for flexural testing at 7, 14, 28 and 56 days were made for every 300 m^3 produced. The entrained air content required was set at 4·5% ± 1·5%.

The final mix design, incorporated into the works, was based on a total cementitious content of 380 kg/m^3. Thirty per cent BS 3892 Part 1 pulverised fuel ash (fly ash) was used in conjunction with Portland cement. Coarse aggregates were single sized limestone 28, 20 and 10 mm. Fine aggregate was zone 2 concreting sand. Water-reducing and air-entraining admixtures were used.

Placing and compacting the pavement quality concrete

A Gomaco 2800 slipform paver was used to lay the PQC (Fig. 5.9). The concrete was transported from the dedicated site batchers to the point of deposition in the Maxon Agitors. Initial spreading in front of the paver was carried out by a rubber-tyred excavator. Compaction was achieved using

Fig. 5.9. *A quality-surface finish was achieved*

24 vibrating pokers mounted on the paver. Following the passage of the conforming plate over the concrete, only minimal finishing by bull float (Fig. 5.10) and hand trowel was required. Surface texture and a curing

Fig. 5.10. *Using fly ash gave a superior quality of surface finish*

Fig. 5.11. Applying the curing agent

membrane were applied by a Wirtgen 850 TCM working directly behind the paver (Fig. 5.11). The completed slab was protected by movable tentage.

Production achieved throughout the works averaged approximately 1350 m^3 per day.

Granular fly ash case study

In August 1997, GFA was used for the reconstruction of a 1 km length of the A52 in Staffordshire, UK. The job consisted of the removal, by planning, of a 400 mm depth of existing pavement. By processing and recycling the plannings, then mixing them with fly ash and lime, a GFA was produced to replace the sub-base and road-base layers of the road. The new pavement consisted of 300 mm of GFA under 100 mm of bituminous surfacing. GFA was chosen for its laying flexibility, immediate stability under traffic, and development of significant stiffness and strength. These attributes were necessary because access to the site was only possible from either end. Thus, any paving material had to be capable of immediate use as well as being able to accommodate the future heavy, slow-moving in service traffic.

Pavement design

The section of road requiring repair is known as Kingsley Bank and is a steep, winding, hilly section of the A52 near Froghall, Staffordshire, UK. The road and surrounding area is geologically unstable. Investigation revealed significant distress in the bituminous layers but a relatively sound and strong formation. In line with the local council policy on recycling and the use of local industrial by-products, it was decided to recycle the existing bituminous layers with lime and fly ash from a local power station.

The design for the reconstruction was based on:

- a formation CBR of 15%
- channelled in service traffic of 8 msa.

The road design was

- 30 mm stone mastic asphalt wearing course
- 70 mm dense bitumen macadam base course
- 150 mm GFA road base
- 150 mm GFA sub-base.

Using material recovered from the road during site investigation, potential mixtures were examined in the laboratory to establish the lime and fly ash contents to satisfy the above. The chosen mixture on a dry basis was 3% CaO (quicklime) + 12% fly ash + 85% planings. It was found that 2% and 4% CaO addition gave virtually identical results. However, 3% was selected for this first application of GFA in the UK. The vibrating hammer optimum moisture content (OMC) for the mixture was 7%. The strength results for 3% CaO are shown in Table 5.8.

Table 5.8. Strength results from granular fly ash used on the A52

Age (days)	7	28	60	90	365
Elastic stiffness: E_{it} (GPa)	–	11·0	–	12·0	13·0
Indirect tensile strength: R_{it} (MPa)	–	0·6	0·8	1·0	1·2
Unconfined compressive strength: R_c (MPa)	1·5	5·0	9·0	11·0	15·0

Production, construction and control

The existing flexible pavement, consisting of 300/400 mm bituminous material on stone/cobbles, was planned out and taken to a nearby Staffs County Council depot, where it was screened into 20–5 mm and <5 mm fractions. Kerbing and drainage work was carried out at the site and the formation tested with the falling weight deflectometer (FWD) and cone penetrometer. This revealed that in places the foundation was weaker than anticipated. Depending on strength and location, these areas were locally excavated to depths of 150 or 350 mm to be reinstated with GFA.

Fig. 5.12. Typical mixing plant for fly ash bound mixtures

At the depot, the screened planings, fly ash, lime and water were mixed in a continuous pug-mill mixer (Fig. 5.12), and the resulting GFA was returned to the site and placed in two 150 mm layers as sub-base and road base. The placing was by conventional paver (Figs 5.13 and 5.14) and compaction by a combination of vibrating and pneumatic-tyred rollers (PTR) (Fig. 5.15). The latter were necessary to produce a tight, crack-free surface and as a test of GFA under traffic . The finished GFA was kept damp by the application of a fine water spray. Under the PTR, the GFA proved its stability under tyres and was able to act as an immediate working platform for access and other subsequent operations. This was important since access was restricted to either end of the job and there was no provision for lorries to turn. This meant that freshly laid GFA was immediately trafficked by lorries bringing in fresh GFA and, later, the surfacing vehicles (Fig. 5.16).

The weather during the GFA operations was variable (see Fig. 5.17). This necessitated tight control of moisture content, particularly in the stockpiles.

On two occasions, over-wet GFA was laid which was impossible to compact. This was rectified by opening up the laid GFA using a toothed JCB bucket. This allowed excess water to evaporate and permitted later compaction. This opening up was possible up to 3 days after mixing. After this time, the GFA began to set and harden.

Fig. 5.13. Laying the granular fly ash road base

Fig. 5.14. Laying the granular fly ash road base (different view)

Fig. 5.15. Compacting the granular fly ash

Fig. 5.16. Finished granular fly ash road base ready for surfacing

Fig. 5.17. Laying granular fly ash on the A52 in the rain

No time restrictions were placed on the surfacing operation. In general, bituminous base course was laid within 1–3 days after the GFA road base, with the SMA wearing course following as required. During the operations, *in situ* compaction was monitored and test specimens were made for strength determination. These confirmed the mix design testing and design assumptions.

The finished product and monitoring

After surfacing but before opening, an FWD survey was carried out. This was repeated in the following spring and on the anniversary of opening. The deflections found are shown in Table 5.9.

Table 5.9. Falling weight deflectometer values

Deflections (mm × 0·001)	Sept. 1997 (13°C)	June 1998 (18°C)	Sept. 1998 (14°C)
Total	170	90	80
Subgrade	40	35	35
Pavement	110	35	20/25

Table 5.10. Core results

E_{it} (GPa)	R_{it} (MPa)	R_c (MPa)
16 (range 11–24)	0·85 (0·6–1·1)	6 (5–7)

At the same time as the 1-year FWD survey in September 1998, 20 no. 150 mm diameter cores were taken. Unusually, all of the cores were removed successfully. The results are shown in Table 5.10.

GFA proved ideal for the A52 work, with its requirements for full flexibility during construction and significant stiffness and strength in the long term. Curing was not necessary, immediate trafficking was possible, and as the FWD and coring exercises illustrate, stiffness and strength have developed as anticipated.

References

1. UK Highways Agency. *Specification for Highways Works, 9C, Cement stabilised fly ash; 9D, lime stabilised cohesive material.* UK Highways Agency, London, 1996.
2. Williams RIT. *Cement-treated pavements.* Elsevier Applied Science, London, 1986.
3. Lister NW, Jones R (Transport Research Laboratory, UK). The behaviour of flexible pavement under moving wheel loads. *Ann Arbor Conference*, Michigan, USA, 1968.

Chapter 6

The use of fly ash for grouting

Introduction

Grouting has been defined as the injection under pressure of suspensions, emulsions and chemical solutions to improve the geotechnical properties of soils and rocks and to facilitate the filling of voids for structural processes.[1] Grout in its simplest terms is a thin fluid mortar. More technically, with reference to ground improvement, grout is a suspension composition used to

- reduce the permeability of the ground
- increase the shearing resistance and subsequent strength of the ground
- fill inaccessible voids.

Grouts may be categorised as chemical, suspension or emulsion systems, and include cement, sand/cement, clay/cement, slag/cement, resins, gypsum/cement, clays, asphalts, bitumens, fly ash, and various colloidal and low-viscosity chemicals.[2]

Grouts are suspension compositions generally produced on site, most commonly comprising fly ash, Portland cement and water. Fly ash has been used for many years as an alternative to sand and cement grouts. Fly ash grouts have a number of important technical, rheological (flow), durability and economic advantages over simple sand and cement grouts.

Types, properties and source selection of fly ash used for grouting

The properties of fly ash which have particular importance in grouts are considered in more detail.

Properties of fly ash when used as a grouting material

Free lime liberated during the hydration of cement is very susceptible to chemical attack and makes little or no contribution to the strength of the mix. Fly ash, having pozzolanic properties, will combine with this lime to produce a stable cementitious material resulting in stronger and more durable grouts. The chemical reaction between lime and fly ash provides a more effective bond than that between sand and cement in weak cement grouts. The pozzolanic activity of fly ash also compensates for the reduction in strength usually associated with fine fillers.[3]

Grading, particle shape, density and moisture content

Fly ash consists of spherical particles with sizes ranging predominantly from 1 to 150 μm (Fig. 6.1).[4] Clearly, fly ash particles are mainly of silt size but there is also a small but significant percentage of finer clay-sized material. The forms of fly ash used for grouting are dry, conditioned and lagoon fly ash. The latter generally is somewhat coarser. These form a continuous grading which imparts the excellent rheological properties to fly ash grout. BS 3892 Part 2[5] and more recently BS 3892 Part 3[6] specify the composition, chemical and physical properties of fly ash for use in cementitious grouts.

Particle size analysis

When choosing a grout for a specific application, the particle size analysis of the suspended particles (and especially the size of the largest) must be

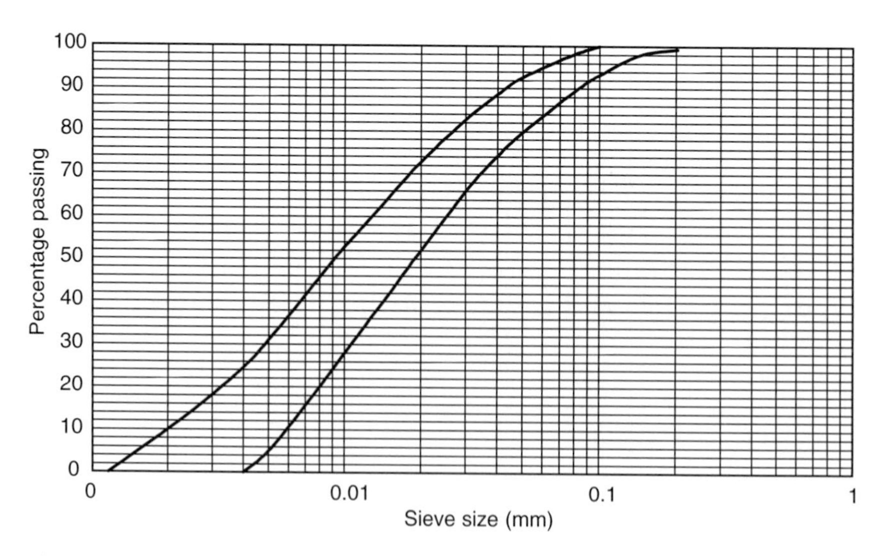

Fig. 6.1. Typical range of grading curves (particle size distribution envelope) for fly ash used in grouting

known. Where several constituents are used, this will be determined from the grading curves for the individual materials. The shape of the grading curve is extremely important in the case of dense grouts, as the size of the largest particles determines the smallest size of passage that can be injected.

Bulk density

The bulk density of compacted fly ash grout is typically 1500–1800 kg/m^3. The particle density of fly ash is approximately two-thirds that of Portland cement and 90% of that of natural aggregates. Typically, fly ash particle densities lie in the range 2·0–2·4. This produces a grout with a significantly lower density than Portland cement and sand grout, which can be beneficial where weight is an important factor. The bulk density of loose (dry) fly ash is about 1000 kg/m^3. Table 6.1[1] indicates the material requirements to produce 1 m^3 of hardened grout together with the associated bulk densities. However, fly ash used for grouting purposes is generally supplied in conditioned form with around 10–15% added water. Thus, the loose bulk density is nearer 1100–1200 kg/m^3.

Table 6.1. Typical designs for grouting mixtures

Fly ash : cement ratio by weight	Quantities (kg) to produce 1 m^3 of grout			Bulk density (kg/m^3)	Approximate compressive strength (MPa) at 28 days
	Fly ash	PC	Water		
1:4	285	1140	500	1925	20–30
1:3	355	1065	500	1920	
1:2	465	930	490	1885	
1:1	675	675	475	1825	
2:1	870	435	455	1760	5–20
3:1	965	320	450	1735	
4:1	1020	255	445	1720	
5:1	1055	210	445	1710	
6:1	1080	180	440	1700	2–5
7:1	1100	155	440	1695	
8:1	1115	140	440	1695	
9:1	1125	125	440	1695	
10:1	1135	115	440	1680	
11:1	1140	105	435	1680	<1–2
12:1	1150	96	435	1680	
13:1	1155	90	435	1680	
14:1	1160	85	435	1680	
15:1	1165	80	435	1680	

PC: Portland cement.

Moisture content

Fly ash can be supplied dry, either in bags or in bulk, or as is more generally the case for large-scale grouting schemes, in a conditioned form, e.g. with between 10% and 15% moisture content.

Properties of fly ash grouts

Water/solids ratios

As the particles which make up fly ash are mostly spherical in shape, fly ash-based grouts have intrinsically enhanced flow (rheological) properties. This means that less added water is required to attain a given flow than for the equivalent sand/cement grout. Water has a significant effect on the properties of grout, both before and after setting. Excessive amounts of water result in increased bleed levels, lower strengths and reduced durability. It is important therefore that the water/solids ratio is kept to a minimum consistent with the flow properties required. In general, the water/solids ratio of fly ash/cement grouts ranges between 0·4 and 0·5 by weight.

Segregation of grout particles (bleeding)

The suspended particles in grouts tend to segregate during flow and settle under gravity when stationary, the latter being analogous to the bleeding of concrete. When bleeding occurs in a grouted medium, fluid-filled channels or spaces are left at the top of the grouted voids. The strength of the grout in the upper parts of the voids is often appreciably less than the average strength. The amount of segregation and bleeding that occurs will depend on such factors as

- the ratio of the volumes of solids and suspending fluid
- the ratio of particle densities of particles and fluid
- the specific surface of the particles
- the rate of strength development of the grout.

Bleeding is measured by the ratio of depth of clear fluid to the original depth of grout expressed as a percentage and is usually plotted as a graph against time. The method of measurement is to perform a simple sedimentation test by filling a 1 l capacity transparent graduated cylinder with the grout. The depth to the grout/fluid interface is then measured at regular intervals. Bleeding will cease when either a stable porosity is reached or the grout starts to set.[3]

The rate of bleeding for a suspension of fly ash in water is compared with that of Portland cement in water in Fig. 6.2,[4] both suspensions containing equal weights of particles and fluid. It can be seen that the rate

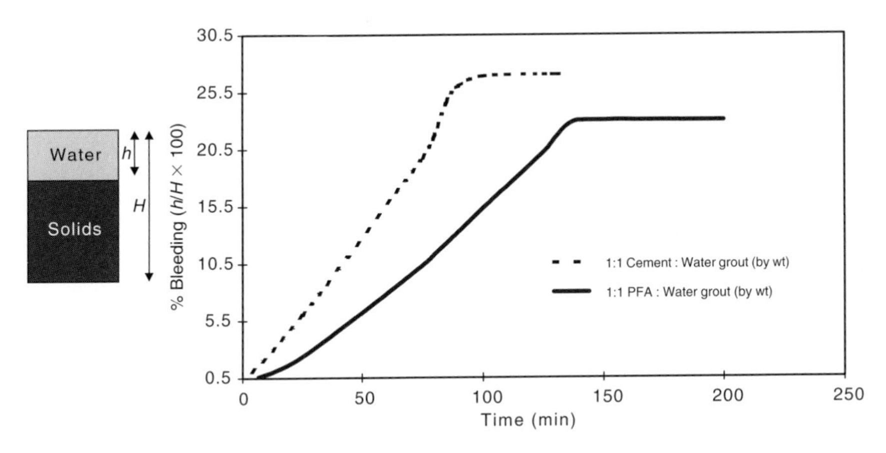

Fig. 6.2. Effect of fly ash (PFA) on bleeding

Table 6.2. Bleed capacity

PFA/fly ash : cement ratio	Water/solids ratio	Wet density (kg/m^3)	Bleed capacity (% of depth with time), 100 mm deep specimens				
			1 h	2 h	3 h	4 h	24 h
1:1	0·40	1765	1	3	3	3	0
	0·45	1715	3	8	6	6	2
	0·50	1677	4	5	5	6	5
2:1	0·40	1718	2	3	3	3	1
	0·45	1679	4	5	5	6	3
	0·50	1648	5	6	7	8	6
3:1	0·40	1695	4	5	6	7	4
	0·45	1650	5	7	9	10	9
	0·50	1628	9	9	10	70	9
5:1	0·40	1675	4	5	5	5	5
	0·45	1641	6	8	8	8	8
	0·50	1599	8	9	10	10	10
7:1	0·40	1680	3	4	4	5	4
	0·45	1611	5	6	8	8	6
	0·50	1587	9	10	11	11	9
10:1	0·40	1643	4	4	5	6	4
	0·45	1620	7	8	8	8	6
	0·50	1575	8	11	11	11	9
15:1	0·40	1658	3	5	6	7	6
	0·45	1608	5	7	9	9	7
	0·50	1582	8	9	10	11	9
20:1	0·40	1645	5	7	8	8	6
	0·45	1607	5	10	9	10	9
	0·50	1580	7	9	10	11	9

of bleeding is lower for the fly ash. This is due in part to the lower specific gravity of the fly ash, the fact that its finer graded particles take longer to settle and its ability to attract and retain water on the particle surface.

In addition, Table 6.2[1] shows the bleed capacity of various fly ash–cement grout mixes over a period of up to 24 h.

Increasing compressive strength

The absence of any coarse aggregate within a grout influences its ultimate attainable compressive strength and therefore the required, or specified, strengths for grouts are lower than those for concrete. This situation reflects the end-use applications and the technical scope of the grouting market. Figure 6.3[1] shows a sample range of 28 day strengths that can be obtained by varying the fly ash to cement ratio and also the overall water/solids ratio. An important point to note is that since grouts are generally specified in terms of attaining a minimum compressive strength, it is possible to make beneficial and economic use of the fact that fly ash–cement grouts exhibit a slower but longer-term gain in strength. This can be achieved by specifying strength criteria at 90 rather than 28 days of age.

Table 6.3 and Fig. 6.4 show the strength development of fly ash–cement grouts typically used for mass filling applications up to 90 days.

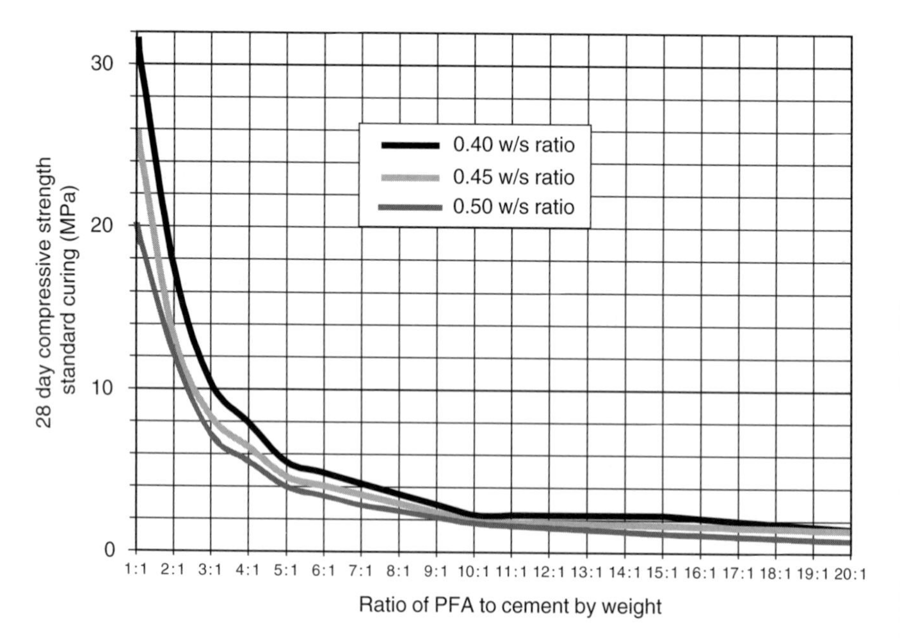

Fig. 6.3. The 28 day compressive strength of low shear mixes: fly ash (PFA) to BS 3892 Part 23

Table 6.3. Compressive strengths with time of grouts

Fly ash : cement ratio	Water/solids ratio	Compressive strength development (N/mm²)			
		7 days	14 days	28 days	90 days
1 : 1	0·40	17·9	22·5	31·7	49·3
	0·45	14·0	18·0	26·0	42·9
	0·50	10·3	14·9	20·0	34·3
2 : 1	0·40	8·7	12·2	17·5	37·9
	0·45	6·2	8·7	13·2	27·7
	0·50	5·4	7·7	12·1	25·1
3 : 1	0·40	4·5	7·0	10·4	23·9
	0·45	3·3	5·3	8·3	18·8
	0·50	2·7	4·4	7·2	19·1
5 : 1	0·40	2·2	3·7	5·1	15·3
	0·45	1·6	2·7	4·4	13·4
	0·50	1·4	2·2	4·0	14·5
7 : 1	0·40	2·2	2·4	5·5	8·8
	0·45	1·6	2·0	4·6	6·5
	0·50	1·4	1·8	2·9	6·0
10 : 1	0·40	1·5	1·9	2·3	4·9
	0·45	0·8	1·4	1·9	4·0
	0·50	0·7	1·2	1·8	2·8
15 : 1	0·40	1·0	1·4	2·3	3·1
	0·45	0·6	1·0	1·7	2·2
	0·50	0·6	0·8	1·2	1·7
20 : 1	0·40	0·7	0·9	1·5	2·0
	0·45	0·5	0·7	1·4	2·0
	0·50	0·5	0·6	0·8	1·3

For example, if the specified design strength required for a grout is a minimum of 4.0 MPa, then

- from Table 6.3[1] it can be seen that if the strength were specified to be attained after 28 days, then a 5 : 1 fly ash : cement mix with up to 0·45 water/solids ratio would be required
- if, however, the strength were specified to be attained after 90 days then a 10 : 1 mix would meet the requirement, thereby halving the cement content.

Strength of the set grout and the soil

The increase in strength of a grouted soil may not bear much relationship to the strength of the grout. The strength developed in the soil can often

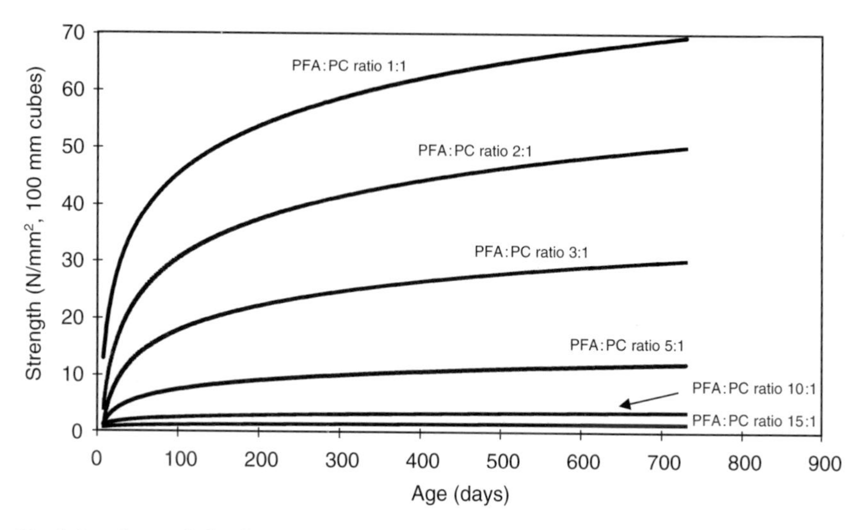

Fig. 6.4. Strength development in fly ash grout (typical strength of grouts: water/solids ratio 0.45). PFA: pulverized fuel ash; PC: Portland cement

depend on the bleeding and segregation properties of the grout. Other important factors may be a lining of soft soil to the surfaces of fissures. When natural ground is injected its bearing capacity will depend on the finer grained soils and very fine cracks. The grout may not penetrate such soils and fine cracks. The chemical reactions that occur when the constituents of the grout, the soil particles and the groundwater come into contact may affect the bearing capacity. For example, free lime in the grout may react preferentially with clay in the soil rather than with the fly ash.

From the above it can be seen that many factors have a great effect on the strength of the grouted ground. Where the bearing pressure or design loads to be imposed are critical factors measurements should be taken. These should be based on a large enough area used as a trial area. The following should be noted:

- Grouts *in situ* are usually fully saturated. Strength tests on samples of unsaturated grouts will give misleadingly high results.
- The pozzolanic reaction between lime and fly ash is very temperature dependent. The *in situ* curing temperature could significantly affect the strength contribution from the grout.

Specification of grout strengths: a note of caution

It is vital that the specification is sufficiently rigorous and detailed as to the suitability of the plant used, together with the mixing method adopted.

Ultimately, confidence in specifying at 90 day rather than 28 day strength is dependent on the control of added water to the mix. Certain computer-controlled batching plants are able repeatedly to attain accurate mixes of grout. In addition, simple batch mixing of calibrated quantities in an agitation tank can achieve an accurate mix. However, confidence is needed that the specified grout strength has a sufficient factor of safety for the specified strength when working with lower cement contents to attain strengths at 90 days rather than 28 days. Here, the ultimate control of water content becomes a critical factor. An excess of as little as 5% water, e.g. from 45% to 50%, can have marked effects on the range of grout strength attainable. Should works proceed based on longer-term gains in grout strength, it is essential that the mix can be controlled very accurately and that a repeatable method with inbuilt safeguards to prevent too much addition of water is adopted. If there is any doubt, it is clearly better to increase the cement content and work within wider parameters.

As stated earlier in this chapter, fly ash is a pozzolana, which reacts with lime to form complex calcium silicate hydrates, which contribute to strength. The lime can be added either as hydrated lime or from the by-product of the Portland cement reaction with water. It is this reaction, which occurs over an extended period (Fig. 6.4),[4] that causes fly ash-based grouts to exhibit slower but longer term gains in strength.

Reduced permeability

Fly ash reduces the permeability of grout through the precipitation of gel products of the pozzolanic reaction. These gels act as a blocking mechanism within the pore structure. Laboratory tests carried out on samples of 32:1 fly ash:cement mix at 0·4 water/solids ratio recorded permeability values of $1{\cdot}3 \times 10^{-8}$ m/s. In practice, increases in density due to consolidation and the use of higher cement contents would be expected to result in lower values.

Void filling properties

The flow properties of a grout must be such that it can be pumped and injected a reasonable distance into the ground, otherwise the number of injection points needed would be unduly large.

Grouts containing a well-graded range of particle sizes are more easily pumped than those containing uniformly sized particles. In addition, rounded particles, such as those in fly ash, can be pumped with less water than angular particles of corresponding size and grading.

The flow properties of fly ash grouts are often discussed in terms of their workability or fluidity. This grout workability is normally measured by a

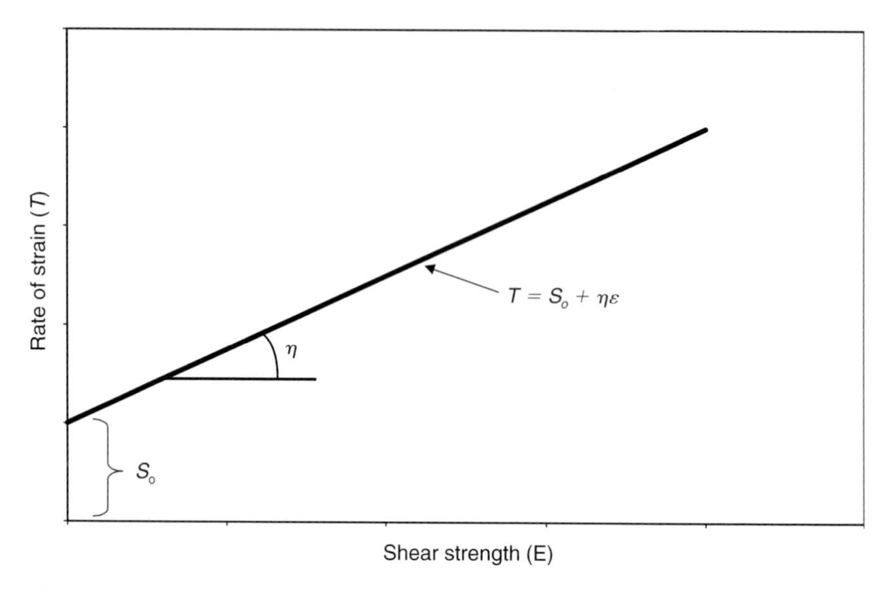

Fig. 6.5. Stress–strain behaviour of grouts if considered a Bingham substance

viscometer using the Bingham fluid characteristics, which are used to describe the observed behaviour of the material.

A Bingham substance is an idealised model material which has both a finite yield stress (S_o) and a strength component that is proportional to the rate of strain (ε). The constant of proportionability (η) is known as the Bingham viscosity of the material. The stress–strain rate behaviour of such a substance is shown in Fig. 6.5 and in the following equation:

$$T = S_o + \eta\varepsilon, \quad (6.1)$$

where T is the shear strength of the material.[2]

Fly ash, because of the nature of its particle shape and size distribution, improves both the shear strength (T) and viscosity (η) values.

In practice, fluidity is measured by use of a Colcrete™ Flowmeter or Marsh cone and is expressed, respectively, as horizontal flow or flow time. Empirical data, from work carried out at the University of Bradford Department of Civil Engineering on flow properties, are shown in Table 6.4.[1] This shows that fly ash grouts with flow values in the region of 450 mm and above are relatively easy to pump through small-diameter pipework.

In addition, fly ash grouts tend to have slower setting times than Portland cement/sand grouts. The overall strength development can, for practical purposes, be considered similar, although at very high fly ash contents this is at a slower rate. Fly ash grouts retain their workability for longer periods after mixing than Portland cement-only grouts (Fig. 6.6).[4]

Table 6.4. Flow properties with time

Fly ash : cement ratio	Water/solids ratio	Time from mixing at 20°C				
		15 min	1 h	2 h	3 h	4 h
1:1	0·40	510	470	400	–	–
	0·45	>700	>700	>700	620	570
	0·50	>700	>700	>700	>700	>700
3:1	0·40	>460	>550	>550	>560	>440
	0·45	>700	>700	>700	>700	>700
	0·50	>700	>700	>700	>700	>700
5:1	0·40	530	530	500	500	470
	0·45	–	>700	–	–	–
	0·50	–	–	–	–	–

Fly ash : cement ratios above 5 : 1 produced flow readings in excess of 700 mm at water/solids ratios of 0.40, 0.45 and 0.50.

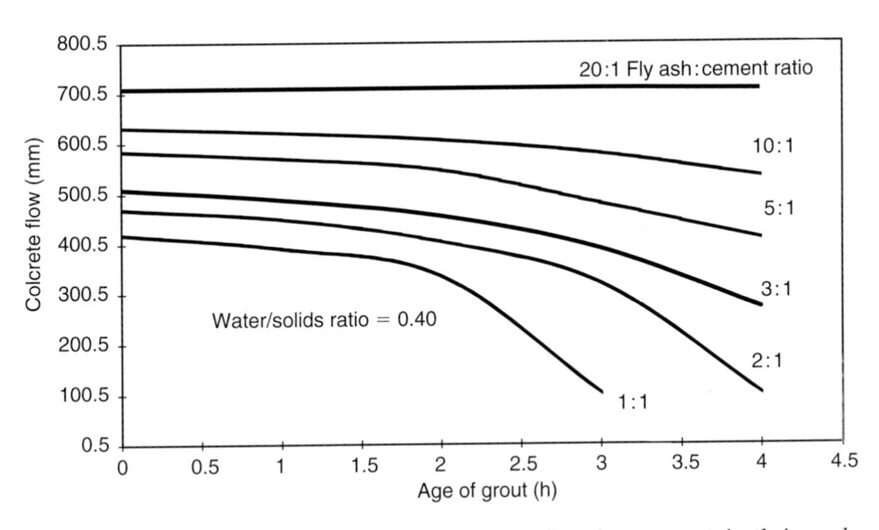

Fig. 6.6. Typical flow properties of various grouts: fly ash grouts retain their workability well

Admixtures

The performance of fly ash grout can be extended by the use of admixtures. These are often specially formulated for grouts to control setting, prevent washout, and improve flow and pumpability characteristics. Retarders, air entrainers and plasticisers are all regularly used with fly ash grouts. In addition, fly ash grouts can be modified to produce little or no shrinkage and attain non-dispersing properties.[4]

The filling of large caverns and open voided ground

Fly ash-based grouts have long been used in operations to stabilise abandoned and collapsing mine workings. This subject has been dealt with at great length by a number of technical papers and publications. One of the most notable of these was *Construction over abandoned mine workings*, published as CIRIA Special Publication 32 in 1984, and co-authored by Healy and Head.[7] This discusses in detail the entire process of using fly ash-based grouts to stabilise both mine shafts and mine workings.

Since the late 1980s and continuing to the present day, several large-scale infilling contracts have been undertaken throughout the Metropolitan Boroughs of Dudley, Sandwell, Walsall and Wolverhampton in the West Midlands (UK), together with Shropshire County Council. Initially, these contracts were funded by the Department of the Environment, and more recently English Partnerships, under the Derelict Land Scheme, to remove the blight of abandoned shallow limestone caverns. The volume of material required to fill these extensive caverns would have been prohibitively expensive using a conventional grout of the proportion 10 parts fly ash to 1 part cement, and therefore alternative materials were considered. After lengthy trials and experimentation it was concluded that a material based on fly ash but with considerably reduced cement content, as little as 2%, was the most favourable. This material, is commonly referred to as fly ash/cement paste owing to its relatively low water/solids ratio. Minimising the water content improves the strength of the grout. This is discussed at length in a number of technical papers, and more specifically by Jarvis and Brooks.[8]

Durability

Sulfates attack ordinary Portland cement grouts by reacting with calcium hydroxide to form gypsum, and calcium aluminates to form ettrignite, resulting in volume increase and disruption. It is well established that fly ash improves the sulfate resistance of grouts by diluting the tricalcium aluminate content of the cement and reducing the calcium hydroxide content as a result of the pozzolanic reaction.[1]

In some circumstances grouts may come into contact with sulfates, e.g. in certain clay soils, contaminated ground or old mine workings. The use of fly ash with Portland cement is normally satisfactory for sulfate resistance. Work at Bradford University showed that in a range of grouts from 1 : 1 to 5 : 1 fly ash : cement grouts, using fly ash from a range of sources, the sulfate resistance improved as the content of fly ash increased. The use of sulfate-resisting Portland cement (SRPC) offers no additional benefits. Fly ash reduces the permeability because of the particle size, shape and the pozzolanic reaction. In addition, shrinkage is significantly reduced compared with Portland cement-only grouts.

The principal benefits of using fly ash in grouts can be summarised as follows:

- reduced bleeding
- increased working life
- improved pumpability and flowability
- reduced permeability
- increased compressive strength and durability
- increased yield per tonne and hence economy
- reduced water/solids ratio.

Types of grout

Fly ash and Portland cement grouts

Since fly ash has approximately the same grain size distribution as cement it is possible to use fly ash as a filler in cement grouts without further limiting the size of cracks or pores that can be injected. Addition of the more rounded fly ash particles to a Portland cement grout improves the flow properties and generally improves pumping and penetration.

By varying the fly ash : cement ratio, dense grouts with a wide range of strengths can be obtained. This makes it possible to choose the most economical proportions giving the desired strength instead of using an expensive, very strong, brittle cement grout for all applications. The ratio of fly ash : Portland cement in common use varies from 1 : 4 to 20 : 1 depending on the strength and elastic properties required. Fly ash/Portland cement grouts do not bleed as much as cement/water grouts but the limitations applied to fly ash/lime mixtures mentioned in the following section may also apply to those of fly ash/cement.

Fly ash and lime grouts

Fly ash/lime grouts are not in general use but tend to be used for specific applications such as fighting spoil heap fires resulting from spontaneous combustion.

Addition of lime to fly ash/water mixtures increases the pozzolanic activity and the final strength of the set grout. Since lime is finer than fly ash and forms a more colloidal suspension the particles within the grout do not segregate so readily and the pumpability is improved. Mortar tests to study the pozzolanic properties of fly ash indicate that the optimum lime : ash ratio for maximum strength is about 1 part of lime to 2·5 parts of fly ash.[9] Fly ash and lime mixes may in time attain strengths comparable with fly ash and cement mixes, but the rate at which the strength increases with time is appreciably lower. At present, there are no data on the strength developed by fly ash and lime grouts, but they will generally have appreciably lower strength than equivalent cement mixes because of their much higher water contents.

Since lime and cement are comparable in cost, there is at present little economic advantage in using lime instead of Portland cement. However, in cases where maximum penetration is required or a slow rate of strength increase may be advantageous, as possibly in tunnel grouting, the use of fly ash/lime mixes may provide the best solution. There are at present no quantitative data available on the bleeding of fly ash/lime/water grouts. Provided the amount of water is kept to a minimum compatible with pumpability, the amount of bleeding is unlikely to be very different from comparable fly ash/Portland cement grouts. Since, however, even very small percentages of bleeding can result in poor adhesion between the grout and the upper faces of cracks, neither type is recommended for grouting fissured rocks where very low permeability occurs or high strengths are required. In these circumstances fly ash/Portland cement/bentonite grouts, as described in a later section, are more applicable.

Fly ash and other combinations of materials

The use of fly ash/Portland cement/bentonite compositions is increasing in popularity, particularly in cut-off trenches. Advantage can be taken of the low permeability of the composite grout, which is particularly useful in preventing the migration of methane gas rising from waste disposal sites.

Advantage can be taken of the self-hardening properties of fly ash to produce low-strength grouts containing only fly ash and sufficient added water to facilitate placing. The strength of a fly ash grout will depend on the residual water content and therefore these grouts should only be used in applications where removal of excess water is possible by either natural or induced drainage.

The unconfined compressive strength of fly ash at 95% compaction (BS 1377 Part 4: Clause 3.3) is shown in Table 6.5[1] up to 90 days of age.

Fly ash and Portland cement/clay grouts

Bleeding of cement grouts can be prevented for the short time required for the cement to develop its initial set. This is achieved by suspending the cement particles in clay suspensions with viscosity and yield strength

Table 6.5. Unconfined compressive strength of fly ash-only grout

Water/solid ratio	Bulk density (kg/m^3)	Unconfined compressive strength (N/mm^2)			
		0 days	7 days	28 days	90 days
0·15	1835	0·08	0·10	0·16	0·19

sufficient to resist the natural fall of the cement particles. Suspensions containing up to 8% by weight of an active bentonite are commonly used for this purpose. The presence of the bentonite makes the grout more viscous, but this may be more than compensated for by the further reduction in segregation and the better pumping and penetration characteristics of the grout. Since the more active bentonite used for this purpose is sodium based, trial mixes should be made to check that the effectiveness of the bentonite as a suspending medium is not reduced by the presence of the calcium ions in the cement or fly ash.

Fly ash/sand grouts

Sand and even gravel can be added to grouts that are being used in situations where the grain size is not a severe limitation. The wider range of particle sizes enables denser grouts to be produced with properties similar to those of good-quality concrete. Fly ash is particularly useful in sanded grouts because of the lubricating action of its rounded particles which makes it possible to pump very dense grouts.

Grouts containing only fly ash

These are essentially cheap, low-strength grouts useful for filling large cavities in the ground. Many fly ash/water mixtures will slurry at a moisture content of about 35% and in this condition the viscosity will probably be too high for pumping. Such mixtures do, however, flow easily at a moisture content of 50%. Provided that the excess water needed to pump the grout can subsequently drain away it is possible to obtain fills with bulk densities very near to those obtained by compacting the fly ash at its optimum moisture content. No strength measurements are available for hydraulically placed fly ash, but data on material compacted at its optimum moisture content have shown that the self-hardening strengths can vary considerably, depending possibly on the amount of free lime present.

The percentage of water should be kept to a minimum compatible with the flow requirements, since excess water carries some of the lime with it as it drains away and thus reduces the self-hardening properties of the grout. Where the excess water is not free to drain away the grout may remain very soft for long periods, and until more data are available on this aspect it is recommended that some Portland cement should be added in these circumstances. Without cement or lime to act as a binder the drying shrinkage of fly ash can be significant.

Grouting techniques

Grouting techniques[10] are many and varied and do not differ significantly when using fly ash within the grout. The following list is not meant to be exhaustive but only to indicate to the reader the techniques that may be used. The use of fly ash in grout, as stated above, improves the properties of the grout.

Permeation grouting

Permeation grouting is the most common type of grouting and is designed to improve the structure of the soil or to control the influx of groundwater. The aim is uniformly to displace water in the voids by the steady outward progression of the grout. Fly ash is preferentially used when open textured soils are encountered, e.g. gravel and coarse sandy soils. For the treatment to be effective, injection pressures must not be so large as to displace the soil particles. Hole positions and depths (Fig. 6.7) are chosen so that the grout from each stage of injection overlaps to form an integrated mass of grouted soil. The sequencing of grouting is arranged so more permeable layers are grouted first.

The ease by which such soils can be grouted depends on the particle size distribution of the soil to be grouted. The cohesive nature and reduced relative water content of fly ash grouts mean they are suitable for most applications. When the soils are of such fine nature, e.g. similar to a coarse sand or fine gravel, that cementitious systems including fly ash grouts cannot fully penetrate them, ultrafine cements and fillers or chemical grouting systems may be used.

Hydrofracture grouting

This relies on injecting cement-based grout at high pressure, up to 4 MPa, to cause localised and controlled fracturing of the soil. The pressures are often higher than the overburden pressure. The method is primarily designed to increase the bearing pressure and shear resistance of the soil. The grout cuts fissures and channels in the soil until it finds voids, which are filled by permeation. It can be used to fill lenses and layers of open-textured soils prior to construction work. It can be used to raise structures before tunnelling operations to compensate for the anticipated settlement.

Fig. 6.7. Grouting injection points

Compaction grouting

This technique involves injecting a stiff grout, 25–50 mm slump, into pipes or casings which are driven into the soil. The grout is pumped under high pressure, up to 7 MPa, to form a bulb-shaped mass. Owing to the relatively low workability of the grout, this tends to compact the surrounding soil rather than permeating the pores. This bulb of grout affects an area of soil that can be up to 20 times greater than the diameter of the bulb. Typically, such mixes will contain 10% Portland cement and fly ash, clay, silt or bentonite. Compaction holes can be positioned either vertically or on the incline, especially when compacting under existing structures. Care has to be taken when working near such structures. Monitoring of movement will be required.

Consolidation grouting

This technique is for grouting of open joints, fissures, bedding planes, faults, cavities, etc., found in rock strata. Consolidation grouting strengthens the rock and reduces the flow of water into the structure. It is often carried out below ground from tunnels and shafts. The defects in the rock strata may result from natural weaknesses or from damage resulting from tunnelling operations.

Curtain grouting

This is used for underground structures and tunnels, normally in rock, which contain some form of liquid or gas. A curtain of grout is formed around the structure, normally radially, to reduce seepage and outflows of the stored material.

Site investigation

No ground treatment can be properly considered until adequate investigation of the relevant ground conditions has been undertaken. The most important part of site investigation is to determine the location, size and type of voids. In rock grouting, an overall geological picture is useful in predicting the location of voids. Geological maps and sections should be prepared to include such information as the boundaries between different geological materials, the inclinations of various layers of soil and rock, contours of the bedrock and locations of prominent faults and falls. In rock grouting, the significant structural features are faults, joints, bedding planes, solution cavities and lava flow structures.

The nature of underground openings may be deduced from groundwater studies, rock exposures, examination of drilled cores, the behaviour of drill tools, core losses, drill water losses, water-pressure tests in drill holes, inspection of large-diameter borings or adits, and finally by the behaviour of injected grout during grouting operations. Surface exposures are invaluable, but care should be taken to remove any superficial debris and to inspect only the intact soil or rocks. Diamond drill cores give useful information on the nature of foundation rocks, but they may yield inadequate or misleading information on openings in the rock. Bedding plane openings are particularly hard to identify unless obviously weathered, because cores tend to break on bedding planes irrespective of whether these are open or not. Even when openings can be identified in rock cores, it is seldom possible to determine the width of the opening from the cores. Drill tools with hydraulic or manual feeds are sensitive to cavities and soft seams, and the behaviour of tools during drilling should be recorded. Core losses may occur for many reasons, but are often due to intense jointing or to the presence of cavities. Areas where core losses are high should be investigated from large borings, inspection shafts or adits. Water pressure tests are extremely valuable and may be done by observing the rise and fall of water in open holes or, preferably, in holes closed by single-or double-packer systems. Tests carried out by injecting water between two packers are particularly useful, since it is possible to obtain values of permeability for the various strata. The testing pressure with water should be roughly the same as that considered safe for grouting, which is ~23 kN/m^2 of pressure per metre of depth.

On all major projects, large man-sized boreholes or shafts should be sunk at strategic positions so that the rock can be inspected visually. Large-scale *in situ* tests should be done on all large projects both before and after grouting to ascertain the effectiveness of the operation. In soft rocks such as chalk, man-sized holes can be drilled down to 30 m using pile-drilling rigs. Borehole cameras can be used in boreholes as small as 75 mm in diameter, and the width of openings can be ascertained accurately. When colour film is used, it is also possible to check grouted areas. Closed circuit television may also be used for inspection purposes in boreholes. It has the advantage that the results can be seen immediately, and interesting zones can be inspected more thoroughly.

The pattern of groundwater flow can be obtained from the location of wells and springs and from pore pressure measurements made in permeable strata. In all but the finest soils, porous cylinders fixed to the bottom of vertical standpipes will provide a reliable means of measuring the pore-water pressure. Whatever type of piezometer is used, it is necessary to ensure that it measures water pressure only in the layer for which measurement is required. It is essential to provide an adequate seal above the measuring point in such a position as to prevent leakage to or from parts of the soil or rock mass which contain water under different piezometric pressures.

Flow rates may be calculated:

$$Q = A \cdot K_w \cdot i, \tag{6.2}$$

where Q = rate of flow (m^3/s), A = total area of pores (m^2) (obtained from cross-sectional area and voids ratio), K_w = permeability (m/s) and i = hydraulic gradient.

When grouting of alluvial deposits is being considered, the first requirement is to have a detailed knowledge of soil profiles and particle size distributions of permeable soils in the various strata. Particle size analyses should be conducted on typical samples of all silt, sand and gravel layers. In the case of silts and fine sands, the analyses should be done on portions of undisturbed samples that have been opened to examine soil structures such as laminations. If laminations are found, then separate analyses should be done on typical material from both coarse and fine layers.

Once an overall picture of the soil strata has been obtained, tests should be conducted to determine the permeability of typical strata. The permeability corresponding to laminar flow through a porous bed is defined by Darcy's law. Approximate values of the permeability can be obtained from particle size distribution data, used in Hazen's formula:

$$K_w = C_1 \cdot D_{10}^2, \tag{6.3}$$

where C_1 is a constant between 10 and 15, and D_{10} is the size (mm) below which 10% of the particles by weight fall. This simple relation is applicable only to uniform sands in a loose state.

If the permeability of the soil is known, more accurate estimates of flow rates can be obtained by a method suggested by Loudon in 1952.[11] In all but small projects, it is desirable to perform some *in situ* permeability tests. The most reliable tests are those in which the quantity of water pumped from a single test well or ring of test wells is used in conjunction with water pressures measured in the surrounding ground. When several layers of different permeabilities are present, it may be necessary to repeat the tests with different penetrations and exposures of the well screens. Even so, except in the case where the various permeable strata are separated by impervious layers, the measured permeability will be partially influenced by the permeabilities of the adjacent layers.

Pumping of grout

Selection of pumping equipment will depend both on the type of application and on the grout mixture. Very large voids can often be filled by gravity flow. Pumps used to grout rock cracks, porous ground and other small cavities should be capable of close control over both pressures and flow rate.

Pumps can be divided into five main types:

- piston pumps
- screw-type pumps
- centrifugal pumps
- pneumatic placers
- flexible tube pumps.

Piston pumps

Piston pumps are capable of pumping grouts at pressures in excess of 1720 kN/m^2. Special valve systems are required for pumping thick suspensions and these may take the form of large ball valves or sliding shutters which cover the inlet and outlet ports at the appropriate times. Ball and piston pumps are suitable for thick mortars at pressures up to 2070 kN/m^2.

Screw-type pumps

Pumps using the Archimedian screw principle are less complicated than piston types but are subjected to heavy wear when used for grouts containing sand. Pressures up to 690 kN/m^2 are common, and some pumps of this type can operate at up to about 1380 kN/m^2.

Centrifugal pumps

These are particularly useful when large volumes of fairly fluid grout have to be delivered at low pressure. Difficulty with cleaning and the vulnerability of the bearings to the abrasive action of grout particles are the main disadvantages of this type of pump. They have been successfully used to pump fly ash slurries, but are not recommended for use with sanded grouts.

Pneumatic placers

Pneumatic placers are the traditional means of grouting tunnel linings. Care must be taken to avoid emptying the compression tank or air will be forced into the previously grouted voids.

Tube pumps

This type embodies a flexible tube along which the grout is forced by external rollers. There are no valves or other moving parts to jam, but the wear on the flexible tube is very severe. Tubes up to 75 mm in diameter are available and give a steady, continuous flow.

Maintenance

For successful pumping, the grout should be kept in continuous movement. Changes in pipe size should be gradual. Good maintenance is essential for efficient grouting and pumps must be flushed out every few hours when cementitious materials are used. Mixers and agitators should be cleaned thoroughly after each period of use, otherwise hardened grout and scale will accumulate which, if dislodged, can cause considerable damage to pumps and lead to blockages in pipelines. Mixed grout, in temporary storage, should be kept moving either by circulating pumps or by agitators in the storage tanks. Auxiliary pumps are often used to transfer the grout from the mixer to one or more pumping stations. All grouts should be passed through suitably sized screens before pumping to remove lumpy material.

Injection of grouts

The injection process is concerned with filling the necessary volume of voids in order to provide the required engineering improvement in the most economical manner. Injection involves the pumping of grout through a distribution pipe system to a tube provided with discharge points through

Fig. 6.8. Drilling the grout injection points

which the grout enters the ground or structure to be treated. The tube must be sealed into the ground (Fig. 6.8) or structure to ensure that the grout travels to where it is required, and does not simply seep back along the pipe to be discharged near the point of entry.

Injection systems

The simplest type of injection point consists merely of a tube with a discharge orifice or perforated length near its end. Simple arrangements of this type are satisfactory for grouting voids in existing structures or in ground immediately adjacent, as e.g. in tunnel grouting. More sophisticated methods are generally required when grouting porous soils or fissured rocks. The main systems used in this type of grouting are as follows.

Single-stage drilling and grouting

This consists of drilling and grouting in one operation along the full length of the drill hole, which penetrates the full depth of ground to be grouted. The grout pipe passes through a packer that seals the top of the hole. This method is limited to formations where the upper soil layers are reasonably uniform. There is no control of the grout injected at the various levels, and the grouting pressures are restricted to the maximum which uplift consideration will allow in the highest part of the strata being grouted.

Multistage drilling and grouting

In this method, a hole is drilled into the upper part of the soil strata to be treated. The surrounding ground is grouted as in the single-stage system, using relatively low pressures in order to avoid uplift of the ground. When this first injection is completed and the grout in the ground has started to set, the grout in the drill hole may be washed out while it is still soft. Alternatively, the hole may be re-drilled through the set grout while drilling for the next deeper stage, which can now be grouted at a higher pressure than was used in the first stage. This process is repeated in stages until the full depth of strata has been grouted. The depth treated in each stage will depend on the variability of the ground and the pressures necessary to inject the grout the required distance. There is much greater control of the quantity of grout injected at each level in the multistage than in the single-stage system.

Pneumatic packers

Additional control can be achieved by using pairs of packers to isolate a particular section of a grout hole, and then injecting this portion through a pipe that passes through the centre of the upper packer. In this way individual seams or fissures can be washed and grouted separately by positioning the packers at different elevations.

Sleeve grout pipe

This uses a pipe with discharge points covered with short rubber sleeves situated at intervals along its length. A hole is drilled to the full depth of the strata to be grouted, and a sleeved pipe of corresponding length inserted. The pipe is sealed along its length with a weak grout. In order to inject through any given discharge point, the particular portion of the pipe is isolated by a double packer. The grout is then pumped through an inner pipe into the space between the two packers. As the pressure increases, the rubber sleeve expands and cracks the surrounding ground. This system has the following advantages:

- Grout can be injected at any selected point, at any stage in the grouting process.
- It is possible to return to a previously grouted level and inject for a second time if this should prove necessary.

Injection operations

Preparation of voids before injection

Loose sediments should be flushed out of cracks before injection using water under pressure. The technique is to drill a number of closely spaced holes and force water or air from one hole to the next with the object of displacing the sediments. In soft, cohesive sediments such as clay it will generally be possible to blow channels through only a part of the clay. Repeated flushing, interspersed with the grouting of the clean parts of the cracks, will usually be necessary. Where thorough removal is essential it may be necessary to make large, man-sized holes to provide direct access to the cracks to be cleaned.

Large surface leaks, which become apparent during water testing, should be caulked before injection to prevent wastage of grout and enable the necessary pressures to be reached.

Trial injections

On large contracts trial injections should be made using the grouts that laboratory tests have indicated as being suitable. The trial should be sufficiently large to enable direct inspection and *in situ* tests to be made in the treated ground. For example, when the object is to reduce leakage, the groundwater flow towards a well or shaft in the grouted area will enable an estimate to be made of the effectiveness of the grouting.

Spacing and positioning of injection points

The basic methods of injection have been dealt with above. Except where pressure washing is required, it is usual to space the initial holes in rock and alluvial ground 12–24 m apart. When these have been grouted, intermediate injection holes are drilled and grouted at half the spacing, and the split spacing technique is continued until the ground will accept no more grout. Final spacing of 3–6 m is common, but the holes may be 1·5 m or less in difficult cases. Care should be taken not to grout too close to previously treated holes until the grout in these has been allowed to set.

When grouting ahead of tunnels, the drill holes are usually fanned out a few degrees so that the ground adjacent to the tunnel is also sealed. It is generally easier to seal the ground around unlined tunnels in this way before excavation than to attempt to grout round the tunnel afterwards. Drill holes in typical applications are 9–15 m long.

Injection of ducts, construction joints, voids inside structural members and around tunnel linings should normally be done from the lowest point and the higher holes successively plugged as grout appears at these levels. In large structures injection pressures of 172–345 kN/m^2 and grout lifts of up to 15 m are common. Slow, steady injection is essential for adequate filling of joints. Rapid injection under high pressure may result in many cavities being bypassed.

Pressure losses in pipelines and in the ground

Whatever system is adopted, the pressure head developed by the pump is partly dissipated in the pipe systems and partly in the porous body being grouted. In both cases there are two components: that required to overcome the shear strength of the grout and the resultant friction along the surface of the pipes or soil particles before flow can commence, and that required to overcome viscous drag during flow. Both components increase as the size of the pipes, soil pores or cracks decreases. The relative importance of pressure losses in the pipes and in the ground depends mainly on the distance the grout is pumped and on the size of the voids that are being injected. For example, when grouting porous gravel or cracks in rocks, the pressure losses in the pipelines are usually very small compared with the losses in the ground around the point of injection. In contrast, when filling cavities around tunnel linings, the losses in the pipelines may be a substantial proportion of the total pressure loss. At the start of an injection the pressures are usually low and gradually build up as the grout advances into the ground from the injection point. The pressures are highest, and exert their influence over the largest zone, at the end of the injection process. The total force produced by an injection can be very large, even when the injection pressures used are in the range of 69–103 kN/m^2, which are relatively low in grouting practice.

Approximate estimates of the pressure losses in the pipelines and in the ground can be calculated from the flow properties of the grouts.

Site control and records

The success of all but the simplest operations depends on the skill of the grouting crews and on detailed supervision by an engineer experienced in the relevant techniques (Figs 6.9 and 6.10). The engineer should ensure that the following aspects are considered.

Prior to the commencement of work, a survey of adjacent structures and services should be made. Further surveys should be made as the work proceeds to check that no damage is being caused by the grouting operations.

Records of injection at each location should be made at regular intervals (about 15 min) throughout the period of injection. The total amount of grout injected at each location and relevant remarks such as bad leakage or tight holes should be recorded. When injection pressures greater than the overburden pressure have to be used, datum pegs should be installed on

Fig. 6.9. Typical fly ash grout mixing plant

Fig. 6.10. Grouting can be cold and wet work

the surface and level records taken at regular intervals to control the amount of ground heave.

Regular inspection and simple field tests should be done to check that the consistency of the grout is within the prescribed limits. An index of fluidity can be quickly obtained by measuring the time taken for the level of the grout in the chamber of a tube viscometer to drop by a fixed amount when using a standard pressure.

Alternatively, simple rotary viscometers can be used. Strength measurements taken on samples of set grout at specified times after mixing provide an additional control on the consistency and composition of the grout.

The effectiveness of the grouting should be checked and for this purpose it will be necessary to take measurements both before and after grouting. It will probably be necessary to form inspection pits or adits in which to conduct *in situ* tests.

References

1. National Power. *Engineering with ash-grout*. Technical Bulletin, National Power PLC, Selby.
2. Somervill SH, Paul MA. *Dictionary of geotechnics*. Butterworths, National Power, London, 1983. ISBN 0-408-00437-1.
3. CEGB. *Fly ash data book, grouting*. CEGB, London, 1969.
4. *Pulverised fuel ash for grouts*. UKQAA Technical Data Sheet No 3, UKQAA, Wolverhampton.
5. BS 3892. *Part 2: Specification for pulverised-fuel ash for use as a Type 1 addition*. BSI, London, 1996. ISBN 0-580-26444-0.
6. BS 3892. *Part 3: Specification for pulverised-fuel ash for use in cementitious grouts*. BSI, London, 1997. ISBN 0-580-27689-9.
7. Healy PR, Head JM. *Construction over abandoned mine workings*. CIRIA Special Publication 32, PSA Civil Engineering Technical Guide 34, 1984. ISBN 086017-218 X.
8. Jarvis ST, Brooks TG. The use of fly ash: cement pastes in the stabilisation of abandoned mineworkings. *Waste Management* 1996; **16**: 135–143.
9. Watt JD, Thorne DJ. *Investigation of the composition, pozzolanic properties and formation of pulverised fuel fly ash*. British Coal Utilisation Research Association Information Circular No. 265, October 1962.
10. Henn RW. *Practical guide to grouting of underground structures*. Thomas Telford, London, 1996. ISBN 0-7844-0140-3.
11. Loudon AG. The computation of permeability from simple soil tests. *Geotechnique* December, 1952.

Chapter 7

Manufactured lightweight aggregates from fly ash

Introduction

There exists a considerable number of lightweight aggregate types. These range from naturally occurring materials from volcanic rocks, e.g. pumice, tuffs, pozzolanas, and volcanic slags, to material produced industrially, e.g. furnace bottom ash (FBA), expanded clay, colliery shale, slate and sintered fly ash. Fly ash/pulverised fuel ash (PFA) has been used in a number of aggregates. The principal aggregate in production is sintered fly ash aggregate, or Lytag, and FBA. Aardelite is a cold fusion process using fly ash and lime that is being marketed at the time of writing, but has yet to go into production within the UK. In the 1980s a material called Taclite[1] was produced using fly ash and FBA using a process similar to the Lytag system. However, it is no longer produced in the UK. FBA is exclusively used in lightweight concrete blocks within the UK and is covered in Chapter 8.

Sintered pulverised fuel ash aggregate

Sintered fly ash aggregate is made from ash that is pelletised, fused and graded. The only material manufactured within the UK of this type, known as Lytag, has been in manufacture since 1960. It is a high-quality aggregate that is used in a variety of applications. Lytag, the company, is the originator of the sintering process. Lytag is an Ash Resources Ltd Company, itself a wholly owned subsidiary of RMC Group plc.

Lytag was originally developed in the UK in the late 1950s. The first production plant, of 100,000 tonnes annual capacity, was commissioned in 1961. Lytag has constructed five factories. Since starting production of Lytag some 15 million tonnes have been produced, saving 30 million tonnes of natural aggregate.

Two further plants were constructed, at Rugeley in 1965 and Tilbury in 1966, giving a production capacity of up to 400,000 tonnes per annum. Product demand increased leading to the commissioning of Lytag's largest UK plant at Eggborough in 1978 (Fig. 7.1), which has a capacity of 250,000 tonnes per annum. During the 1980s, a Lytag factory in The Netherlands for BV VASIM, a subsidiary of a major electricity utility, was built. This plant produces 150,000 tonnes per annum of Lytag. A similar sized plant was also commissioned in Poland in 1990.

For power stations (Fig. 7.2) it has been found that constructing fully engineered disposal sites requires substantial capital expenditure programmes and there are high operating costs to maintain environmental standards.

Lytag manufacture utilises fly ash, thus decreasing the extraction of local natural aggregate resources and preserving them for future generations.

Fig. 7.1. Eggborough Lytag plant

Fig. 7.2. Fly ash for Lytag is sourced from a UK coal-fired power station

Lytag has been used regularly throughout Europe and in many countries world-wide. The benefits of using Lytag are:

- it utilises a by-product material
- finite natural aggregate resources are preserved for future generations,
- as Lytag is an insulating material, when used in the construction of buildings it can reduce heating requirements
- when used in concrete the reduced mass of the resulting structure leads to a reduction in foundation capacity and hence a lower volume of concrete is required
- it has a variety of applications, including horticulture, concrete, water-treatment filters, vehicle arrestors and children's play areas.

Producing Lytag from fly ash

The Lytag pelletising process is sensitive to the fly ash particle size grading and loss on ignition (LOI) and new sources are assessed for suitability through a pilot plant. Fly ash contains a small percentage of unburnt fuel, which is measured by the LOI test and the results are reported as a percentage. Ideally, the LOI of the fly ash should be ~6%, although variations are acceptable. Fly ash conditioned or lagooned with fresh water at residual moisture content levels of <14% is acceptable subject to appropriate handling facilities.

Supplementary additions may be blended with the fly ash to bring benefits to the process and the product. Pulverised coal can be included where fly ash LOI contents are low to aid the Lytag sintering process. Minor additions of waste materials such as cement kiln dust (CKD) and incinerated sewage sludge ash (ISSA) can be included.

The Lytag production process

Lytag is formed because of the agglomeration which occurs as the fly ash particles fuse and bond together. The transformation of fly ash to aggregate has long been established; however, the Lytag process brought a new technique to fly ash conversion (Fig. 7.3). The process required that no binding agents were used and that no pre-heat was applied to the pellets before ignition. The only additives that may be needed are water and additional fuel.

Water assists with the binding of particles. The fuel, typically in the form of high carbon fly ash or coal, is needed to ensure that the homogeneous pellets contain sufficient energy to raise the temperature of the fly ash to the sintering point later in the process.

The Lytag production process is, by necessity, a fully automated system using advanced process and quality-control techniques. The unprocessed dry fly ash is pneumatically conveyed from the power station into suitably sized storage silos to meet process needs and fly ash availability; however, the process is also capable of processing stockpiled moistened or 'conditioned' fly ash. The transfer of fly ash is carried out in fully enclosed

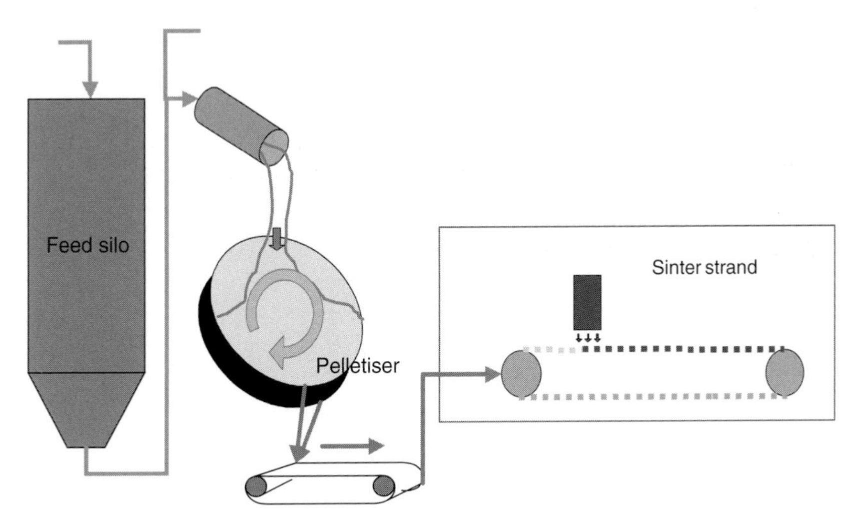

Fig. 7.3. Schematic layout of a production facility

systems to prevent dust emissions. Excess air displaced by the fly ash when being conveyed and fluidised is extracted from the system by dust extraction filters.

Fly ash is fed from the storage silos via variable-speed rotary feeders to process continuous screw mixers where a metered amount of water is added to achieve a moisture content of ~14% (Fig. 7.4). The mixers are fully enclosed and receive metered quantities of fly ash and supplementary materials. The damp fly ash discharged from the mixers is transferred by belt conveyors to inclined, rotating pan pelletisers as shown in Fig. 7.5. The damp fly ash is formed into pellets by the further addition of metered process water sprayed on to the rotating pan until the required size of pellets is achieved. The full-size 'green pellets' are discharged from the bottom rim of the pelletiser pan on to a belt conveyor and transferred to the sinter machine feed hopper (Fig. 7.6). The green pellets are evenly laid across slowly moving grate pallets of the sinter machine (Fig. 7.7). The pallets pass under an ignition hood (Fig. 7.8), fired by waste oil or gas at the driven end of the sinter machine. The top surface of the green pellet bed automatically ignites as the temperature in the hood is maintained at 1000–1200°C.

A down-draught is applied to the underside of the pellet bed through a series of ducts running the total length of the sinter machine (Fig. 7.9).

Fig. 7.4. Conditioning the fly ash with water before pelletising

Fig. 7.5. Pelletising the fly ash on rotating pans

Fig. 7.6. Discharge of the 'green pellets' from the pelletising pan

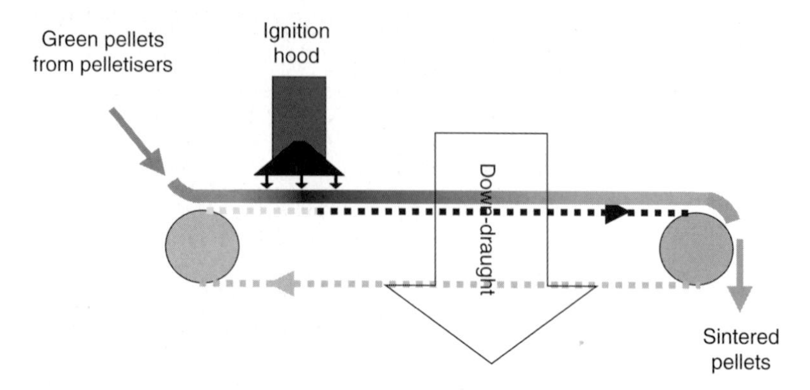

Fig. 7.7. Schematic of sinter strand

Fig. 7.8. Ignition hood

As the slowly moving pallets emerge from under the ignition hood, the burning zone is slowly drawn through the bed depth of the green pellets. The full depth is burnt through by the time the pallet discharges the finished product at the non-drive end of the sinter machine. As the finished product slides off the pallet at the discharge end, it falls into a slowly rotating finger breaker unit that breaks any agglomerated material created

Fig. 7.9. *Pellets being carried along the sinter strand*

during the sintering process. The down-draught air for the sintering process is provided by a fan pulling ~3400 m^3/min. The air drawn through the bed via the ducts is fed to a chimney via an electrostatic precipitator to remove remaining particulate material.

The sintered pellets are directed to an enclosed screening unit to separate the Lytag into the two main products, granular (>4 mm) and fines (<4 mm). The system allows a variety of sizes to be produced to meet market requirements. The granular material is discharged via a conveyor into a collection pit via a water spray curtain, installed to minimise dust emissions and cool the material (Fig. 7.10).

Stringent quality-control and process test measures are used on both raw material and the finished product to minimise waste and to achieve a consistent and reliable Lytag product. Control of the raw material includes measurement of moisture content, particle size distribution, LOI and principal oxides. The product is tested for strength, size grading and density, and its performance in concrete is assessed.

Properties of Lytag

The pelletising process provides a rounded spherical shape to the finished aggregate (Fig. 7.11). The internal structure is a honeycomb of generally interconnected voids of varying size and shape amounting to some 40% of

Fig. 7.10. Processed Lytag cooling before grading

Fig. 7.11. The final product: a sintered lightweight aggregate

the volume. This void space provides a reservoir for absorbed water and the expansion chambers needed to avoid the disruptive effects of ice formation if the aggregate is exposed to freezing. The air voids also act as an insulating material enhancing the thermal properties and frost-resisting characteristics of the finished concrete. As the surface consists of open pores, moisture can enter and exit the aggregate, which helps fully to hydrate the cement paste so that better quality concrete may be produced. The chemical composition of Lytag is similar to the raw materials used to make the green pellets, with the exception of the carbon content, which is reduced significantly during the sintering process.

The bulk density of the Lytag granular is typically 825 and 1100 kg/m^3 for the fines. Density can vary with a variable LOI percentage in the raw material, although process control measures and/or the inclusion of additional supplementary materials can be used to minimise the effect. Boiler innovations and the use of coal from multiple sources will influence most fly ash characteristics but, generally, fly ash from a single source will be relatively consistent.

Although Lytag aggregate absorbs water, this can greatly assist in the long-term hydration of the cement paste, resulting in high-quality, high-strength concrete. In addition, Lytag does not degrade when exposed to the atmosphere, even under freezing conditions.

It is practically and commercially unwise to attempt the production of Lytag using an unknown fly ash in a full-scale factory without first commissioning a technical feasibility study including pilot plant trials

Applications of Lytag

Lytag is a very strong, lightweight aggregate that has proven durability and a record of acceptance internationally. Lytag is purchased by many different market sectors, from the construction industry, which includes its use in readymix concrete up to 100 N/mm^2, as indicated in Table 7.1 to water companies, which use Lytag as a filter medium.

Table 7.1. Properties of various aggregate types

Material	Concrete strengths (MPa)	Density of concrete (kg/m^3)	24 h water absorption of basic aggregate (% by volume)
Lytag	5–100	1550–1900	15%
Slag/furnace	1–10	N/A	N/A
Bottom ash	5–100	2300–2500	Variable, depending on type of aggregate
Natural aggregates			

NA: not applicable.

As well as being lightweight, Lytag has other useful features. As it is an insulating material, it can reduce future energy demand when used in buildings. In the UK, many houses and buildings are constructed using concrete blocks incorporating Lytag. The material is chemically inert and does not contribute to alkali silica reactions in concrete. Lytag therefore has many applications, which include the following.

Concrete blocks

A block made of Lytag tends to have a higher strength than other lightweight blocks (strengths up to 14 MPa are easily obtained from blocks weighting 20–25% less than normal-weight blocks). The lightweight nature of the blocks leads to transport cost savings and reduces manual handling problems. Lytag has excellent insulating properties, so concrete products using Lytag conduct less heat. When used in buildings this provides a more comfortable, warmer environment for similar energy input than if standard concrete is used.

Concrete

Lightweight concrete can be manufactured at densities between 1550 and 1900 kg/m^3, compared with normal-weight concrete of 2400 kg/m^3. This allows less material to be used in construction. Slimmer structural elements can be achieved and smaller foundations with less piling are needed as the dead load is reduced by about 20%. As spherical aggregate, Lytag produces workable concrete (Fig. 7.12). This speeds up the construction process as the material flows around reinforcement easily and ensures a lower risk of poor workmanship. Better quality work should result. Major projects in which Lytag concrete has been used include:

- Canary Wharf Tower, London, UK (Fig. 7.13)
- bridge across the River Rhine, Arnhem, The Netherlands
- Hunterston Oil Platform for BP Oil, Scotland, UK
- Commerz Bank Tower, Frankfurt, Germany.

Screeds

A Lytag screed is a cement-bonded Lytag lightweight aggregate no-fines base coat, with a 4 : 1 by weight sand: cement topping. This is used to provide a smooth, insulated finished floor or roof level. The no-fines base coat, of either 10 : 1, 8 : 1 or 6 : 1 by volume, can be bonded to the sub-base or unbonded, or float over a further insulating layer.

A Lytag screed has a density approximately one-half that of a sand/cement screed, so substantial weight savings can be gained. Not only can these savings in weight lead to lower construction costs, but a Lytag screed

Fig. 7.12. Lytag concrete

Fig. 7.13. Canary Wharf tower, London, UK

uses only one-half the cement of a normal-weight screed, again saving money. Owing to the regular shape of Lytag, air can easily pass through the base coat, so more rapid drying out occurs. This allows for speedier access on to the floor, to continue with construction operations.

The extremely low shrinkage characteristics of a Lytag screed allow large bays to be poured. There is no restriction on the maximum depth that can be laid, so falls can be achieved easily. Additional material can be added for thermal insulation.

Refractory concrete

Lytag aggregates have low densities and high strengths. These properties, combined with inherent fire resistance and insulating characteristics, make an ideal aggregate for use in refractory concrete products, such as flue and chimney linings.

Drainage

Because of its spherical shape, Lytag is an excellent material to use in land drainage (Fig. 7.14). It flows easily into trenches, so time is saved in placing. Lytag virtually eliminates future settlement since it moves readily to fill all voids when it is being placed. Lytag allows about six times more water to pass through it than do ordinary aggregates, thus draining the land far more quickly.

Fig. 7.14. Lytag used as a lightweight fill material.

Vehicle arrestor beds

A Lytag arrestor is a prepared bed of Lytag pellets used to form escape routes on roads which have either steep gradients or dangerous bends; on motor racing tracks and at airports. Owing to Lytag's regular shape and strength, it decelerates vehicles of all sizes, at a controlled rate, with minimal damage to the occupants or vehicle.

In the UK, arrestor beds have been in place for up to 25 years with no degradation of the material. Unlike natural aggregates, Lytag does not compact after being placed, so minimal maintenance is required to keep the arrestor bed functioning and ready for use.

Water filters

The properties of Lytag, as an artificial material, are closely controlled during the production process. This, combined with its resistance to wear and degradation, makes it an ideal material for use in water filtration for sewage systems ('Aqualyt').

Lytag has numerous applications and the development of new opportunities continues at a pace. Currently, predominant market is in concrete, where its demand is determined by the technical and environmental benefits that it uniquely provides.

The Aardelite process

This is a cold-bonded process for the production of lightweight aggregate. The process combines fly ash, calcium oxide (quicklime), additives, fine sand and/or FBA. A mixture of fly ash and sand is agglomerated with calcium hydroxide (hydrated lime) as a binder to form green pellets. These are formed using pelletisers, similar to those used in the Lytag process. These pellets are then bound in a mixture of calcium oxide (quicklime) and sand. The mixture is moistened and heated by low-pressure steam up to 80°C. This moisture and heat produces an exothermic reaction raising the temperature to around 96°C. Calcium silicate hydrates and aluminates are formed in the hardening process, which takes up to 8 h. On completion, pellets are screened to size prior to marketing. The same principles can be applied to other materials, such as phosphate residues and incinerator waste ashes.

The composition of the Aardelite pellet production is 48% fly ash, 45% sand, 4.5% lime, and 2.5% additives and water. The Bulk density of Aardelite varies from 980 to 1020 kg/m^3. The weight of concrete produced with Aardelite gravel is 18–20% lower than concrete produced from natural aggregates.[2]

Aardelite can be used in a wide range of applications, including blocks, concrete, precast elements, asphalt for roads and paving stones. One plant in the USA[3] has produced over 1,000,000 tonnes of material since opening in 1988, with a fly ash utilisation of 95%. However, at the time of writing there is no Aardelite plant in the UK.

References

1. Concrete Society. *Lightweight concrete.* Construction Society, Crowthorne 1980. ISBN 0-86095-861-2.
2. Boas A, Spanjer JJ. *The manufacture and the use of artificial aggregates from fly ash produced according to the Dutch cold bonded 'Aardelite' process.* CEGB, AshTech' 84, 1984: 577–582.
3. *Ash Management brochure.* Progress Materials Inc., Florida.

Chapter 8

Fly ash in aerated concrete blocks and furnace bottom ash in lightweight concrete blocks

Introduction

In 1999, the UK market for pre-cast concrete masonry units was ~8.6 million m^3 (DETR). The vast majority of precast concrete masonry blocks produced in the UK using modern technology and quality-controlled casting methods contain fly ash or furnace bottom ash (FBA). Fly ash may be included as part of the cementitious material, as an aggregate or both to produce dense and aerated concrete blocks. At least 30% of the cement content may be replaced with fly ash.

Of this total market ~24% were lightweight aggregate blocks in the density range of 1000–1500 kg/m^3 and 31% were autoclaved aerated concrete (AAC) blocks in the range of 400–800 kg/m^3. These blocks have high levels of thermal insulation and a high strength/weight ratio, and are able to meet acoustic and fire insulation requirements. The blocks, which conform to UK standards, contribute overall costs savings arising from a number of secondary savings. Lighter foundations and structural frames and the need for less insulation all produce real benefits. They are easily cut, worked and laid, with minimum maintenance and low handling costs.

Autoclaved aerated concrete

AAC, also known as 'Aircrete' in the UK, is a lightweight building product used in the construction of domestic dwellings and commercial buildings. It is manufactured as blocks (Fig. 8.1) or steel-reinforced panels. Characterised by its fine cellular structure, with air pores ranging from 0·1 to 2 mm, AAC has a high ratio of compressive strength to density. This property allows AAC to be used as a load-bearing unit where efficient thermal insulation is required.

Fig. 8.1. Typical autoclaved aerated concrete blocks produced with fly ash

AAC is made by reacting together finely divided calcareous and siliceous raw materials in saturated steam at temperatures above 100°C (hydrothermal conditions). Steam curing of specimens within pressure vessels (autoclaves) at several times atmospheric pressure ensures that hydrothermal conditions are maintained. The curing process is termed autoclaving. The calcareous raw material is normally quicklime (calcium oxide) or a combination of quicklime and Portland cement. The siliceous component is most often finely divided quartz, obtained from sand or sandstone. Alternatively, a raw material containing amorphous silica, or alumino-silicate glass, may be used. Fly ash from coal-burning power stations is a suitable siliceous raw material and is extensively used within the UK for the manufacture of AAC.

In principle, the cellular structure of AAC can be formed either by gas evolution from a chemical process or by mechanically entrained air. However, addition of fine aluminium flakes to a cementitious slurry is the only common industrial process used. Typically, most cells are discrete and do not connect with adjacent cells. This ensures low moisture permeability through AAC, despite its cellular structure.

The intercellular matrix within AAC is bound together by the reaction products formed during autoclaving, either calcium silicate hydrates or a combination of calcium silicate hydrates and calcium alumino-silicate hydrates. These reaction products are more crystalline than analogous binding phases formed in conventional concrete or mortar.

History

Experiments with autoclaved concrete were conducted in the nineteenth century. Patents relating to autoclaved sandlime bricks were filed in Britain and Germany in 1866 and 1888, respectively. Industrial manufacture of sandlime bricks started in Germany in 1894. Cellular concrete has also been produced since the nineteenth century. Johan Axel Ericksson, a Swedish architect, successfully combined these technologies and in 1924 developed the method for producing AAC. Full-scale production of AAC began in Sweden during the late 1920s and it rapidly became an established building material throughout Scandinavia. After World War II, manufacture started in other European countries, as well as Asia, Latin America and the Middle East. AAC was introduced to the UK in 1951.

North America is the only industrialised region that has not manufactured AAC in significant amounts. Readily available timber throughout Canada and the USA has been an explanation for the slow growth of AAC use. However, dwindling timber reserves have led to increased costs for construction lumber, making AAC a financially more attractive building material. Consequently, AAC is gaining acceptance within North America and three full-scale manufacturing plants established in the south-eastern USA use ground quartz sand as the siliceous raw material.

Fly ash use in autoclaved aerated concrete manufacture

World-wide, most AAC is manufactured using quartz sand as the siliceous raw material, rather than fly ash. Quartz sand is used throughout continental Europe, Japan, Latin America and the Middle East. However, the use of fly ash is well established in the UK, the former Soviet Union, China and India. Most AAC made in the UK uses fly ash as a siliceous raw material.

The variability of fly ash must be considered during AAC manufacture. However, there are several advantages associated with the use of fly ash for the manufacture of AAC. Environmental benefits are achieved by using a waste raw material as an alternative to a primary aggregate such as sand. The autoclaved matrix that results from the use of fly ash, because of the influence of aluminium ions, has a high resistance to sulfate attack. A low thermal conductivity can be achieved for AAC made with fly ash, owing to the low conductivity of the fly ash.

The environmental benefits of using fly ash are of particular interest in North America. As disposal costs for fly ash sent to landfill or lagoons increase, AAC manufacture becomes an attractive outlet for the by-product.

An EPRI-sponsored project in 1993 used a mobile pilot plant to demonstrate the feasibility of using fly ash from several power stations for the manufacture of AAC. A full-scale factory is being built at the Bull Run Plant near Clinton, Tennessee.

Manufacture of autoclaved aerated concrete

Manufacture of AAC is unlike that of conventional concrete or mortar. The raw materials are fine powders, without aggregate particles of any significant size, and the starting point is a water-based slurry. Three aspects of the process differentiate it from other concrete pre-casting methods. There is an initial 'aeration' stage in which the slurry expands to form a stable cellular mass. Once stiffening of the mix has occurred and sufficient 'green strength' has been achieved, the cellular mass is cut into individual masonry units. Finally, autoclaving at elevated temperatures promotes hydrothermal reactions, which form a stable high-strength intercellular matrix. The manufacturing process is detailed below.

Weigh batching and mixing

The dry raw materials are transported from storage silos into weigh bins above the batch mixer. Mix water and any slurried material are weighed or volumetrically dosed into the batch mixer.

Casting

A low-viscosity water-based slurry (~10 Pa·s), containing fly ash, cement, quicklime and additives, is produced in the mixer. Finely divided aluminium powder is the final raw material added to the mix. A total mixing time of 2–4 min is typical, but the aluminium powder is dispersed for only 10–20 s before the slurry is discharged from the mixer into oiled rectangular steel moulds. Typically, the moulds are 0·5–0·7 m deep and have a volume of 3–4 m^3. The mass of dry raw materials used in a single mix of AAC ranges from 1500 to 3000 kg, depending on the final density of the product. An AAC mix is normally termed a cake once it has been cast (Fig. 8.2).

Contact with the alkaline mix (pH 9–12) leads to the reaction of the aluminium flakes, liberating hydrogen gas as

$$2Al + Ca(OH)_2 + 2H_2O = Ca(AlO_2)_2 + 3H_2. \quad (8.1)$$

After casting, numerous small bubbles of hydrogen (0·2–2 mm) form within the slurry. Approximately 1·2 m^3 of hydrogen gas is generated from every kilogram of aluminium powder used. The cake expands rapidly within 15–30 min, by up to 220% of its original volume. The volume of hydrogen gas generated within the mix is proportional to the amount of

Fig. 8.2. Casting an autoclaved aerated concrete cake

aluminium powder used. Therefore, a certain quantity of aluminium powder is necessary in order to achieve the required expansion and specified density. Too little aluminium powder will result in a low cake of high density, whereas too much gives a high cake of low density.

The cement and quicklime within the cake react exothermically with the water. Gels of calcium silicate and calcium aluminate hydrates form because of the hydration of cement grains. The quicklime slakes to gelatinous calcium hydroxide (Equation 8.2), which has a strong dewatering effect on the mix:

$$CaO + H_2O = Ca(OH)_2. \tag{8.2}$$

The changes increase viscosity and stiffen the cake. A rise in internal temperature takes place, from 40–45°C at casting to 65–85°C after about 120 min. The hydrogen trapped within the cells diffuses out, to be replaced by air. At this stage the cellular mass has gained sufficient green strength to allow demoulding, handling and cutting into individual blocks.

For a particular formulation, using specific raw materials, the critical control parameters at casting are as follows:

- cast temperature
- water content
- amount of aluminium powder.

The amount of quicklime in the formulation and the temperature of the mix water largely determine cast temperature. In general, cast temperatures must be controlled between 40°C and 45°C. Lower temperatures are likely to give slow cake rise and prolonged set. However, excessive cast temperatures are to be avoided, since this results in rapid expansion of the cake to give a coarse, unattractive cellular structure.

Maintenance of the correct water content of the slurry is important to control the early stage of the manufacturing process. Water/solids (W/S) ratios between 0·40 and 0·55 are typical for AAC mixes. If the W/S ratio is low the mix will be very viscous at casting. Expansion of the mix may be impeded and the cakes will not reach the correct final height. In addition, significant amounts of air may become entrained during mixing or at discharge into the mould. Large air bubbles retained in the expanding mix will be evident on the surface of the finished AAC blocks.

Conversely, if the W/S content of the slurry is too high, the cast AAC mix will have a low viscosity and the stability of cake may be reduced. This is critical in the first few minutes after casting, when the cake is expanding rapidly. There is a fine balance between the cohesiveness of the mix and the pressure exerted by the hydrogen gas generated from the reacting aluminium flakes. If the expanding mix has low viscosity and is not sufficiently cohesive the hydrogen bubbles will coalesce to form large cells in the weak cementitious mass. This coalescence can be so extensive that the cellular structure within the cake breaks down completely. Cake stability is critical for the successful manufacture of AAC and is affected by raw material selection and the formulation used.

Setting

Reaction of the aluminium powder is normally complete 15–30 min after casting, depending on the formulation and choice of aluminium powder. The cementitious mass has sufficient green strength to maintain a stable cellular structure, but is still too soft to cut. Hydration processes, initiated at mixing, continue to stiffen the cake. Heat liberated from the exothermic reactions of the quicklime and cement raises the internal temperature from ~45°C to ~80°C within 120–150 min after casting. The cellular structure at the exterior of the cake insulates the interior and aids the temperature rise. Cakes are normally allowed to stiffen within tunnels or enclosed bays for 90–180 min. These structures normally can be heated to accelerate setting. Once sufficient green strength has been achieved (~0·5 MPa), demoulding occurs and the cake is moved on to the cutting stage.

Common industrial practice is to rotate the mould through 90° along its main axis in order to remove the AAC cake (Fig. 8.3). One side of the mould is also unlatched during rotation so that the AAC cake is supported underneath. Thus, one of the sides of the mould becomes a base, which supports the cake during the rest of the manufacturing process.

Fig. 8.3. Autoclaved aerated concrete cake rotated through 90° before demoulding

Fig. 8.4. Demoulded autoclaved aerated concrete cake about to enter the cutting line

Cutting

The stiffened cake (Fig. 8.4) must be cut into individual blocks or panels before curing within steam autoclaves. Cutting is normally achieved by three sets of thin steel wires. The cutting line mechanism moves the

AAC cake horizontally through the three sets of wires. At the first cutting stage vertical wires remove excess material from what were the 'top' and 'bottom' relative to the cake rise. A second set of reciprocating horizontal wires cuts the cake into a series of slices. Finally, the cake is held stationary while an array of reciprocating horizontal wires cuts down through the cake.

Block dimensions are determined by the spacing of the wires on the cutting line. In the UK, tolerances on dimensions for building blocks are specified in BS 6073.[1] The most common block size is 440 mm long by 215 mm high but other sizes are available. Block thickness is varied, depending on the thermal insulation characteristics and load-bearing properties that are required. This dimension is achieved at the second cutting stage and therefore the mechanism allows the spacing of these reciprocating wires to be changed readily (Fig. 8.5).

Some off-cut material is produced at the cutting line. Since this contains expensive raw materials, such as lime and cement, it is slurried and recycled to the mixer as an ingredient. This waste slurry has the additional benefit of stabilising the AAC cake during the first few minutes after casting. It is believed that this effect is due to the seeding influence of fine grains of hydrating calcium silicate hydrates and calcium hydroxide.

Fig. 8.5. Autoclaved aerated concrete cake passing through the horizontal wires of the cutting line

The green strength developed within the AAC cake is sufficient to enable demoulding, cutting and loading of uncured material into the autoclaves. The time from casting to autoclaving is typically 4 h or less. If the cakes were not autoclaved, the hydration reactions would continue, giving increased strength. However, even with prolonged curing for several days or weeks, the strength gain is insufficient for the structural strengths required. Satisfactory compressive and tensile strengths for constructing load-bearing walls can only be achieved by curing in steam autoclaves.

Autoclaving

Aerated concrete cannot be cured by dry heat alone. This would drive water of hydration from the cellular structure and prevent the formation of calcium silicate hydrates and calcium aluminosilicate hydrates. Autoclaving is essential because the combination of elevated temperatures and moisture ensures that hydrothermal reactions and the rapid formation of semi-crystalline and crystalline reaction products occur. The four stages of autoclaving are reviewed below.

Purging

Saturated steam conditions are required throughout the autoclaving process. This requires the efficient removal of air from the sealed vessel. Air/steam mixtures must be avoided during the autoclaving process. At saturated steam conditions, without air contamination, there is a well-defined relationship between the pressure and temperature. However, even a relatively modest amount of air will give a lower temperature than predicted from the steam curve. Air pockets can also form within an autoclave, or a thin film of air can adhere to the surface of cakes. Since air is an effective thermal insulator, poor heat transfer can arise which adversely affects the autoclaving process. In the purging stage, air is removed from the autoclave by evacuation, or passing steam marginally above atmospheric pressure through the vessel.

Pressurisation

The autoclaves are pressurised at a controlled rate to the desired maximum pressure. This ensures that the AAC cakes do not experience thermal shock. In addition, there will be an economic pressurisation rate, depending on the steam plant available. It is normally desirable to use 'blowover' steam, whereby an autoclave is partially pressured with steam from another vessel which is at the end of its dwell cycle.

Dwell

The dwell stage is the longest part of the autoclave cycle. The autoclave and its contents are maintained at a set pressure for a period of 6–12 h.

Depressurisation

The final stage of the curing process is the controlled depressurisation of the autoclave to atmospheric pressure, minimising the risk of thermal shock.

Figure 8.6 shows AAC cakes being unloaded from an autoclave.

Hydrothermal reactions

Calcium silicate hydrates (CSH) are considered to be the reaction products that make the largest contribution to the development of strength within autoclaved materials. Calcium silicate hydrates form by the reaction of calcium ions and solubilised silica to give insoluble products that are generally poorly crystallised within conventional concrete. Autoclaving accelerates the process and results in further crystallisation of initial CSH phases. The concentration of both calcium ions and solubilised silica within the aqueous phase determines whether saturation occurs and the calcium to silica ratio of initial CSH phases.

Fig. 8.6. Autoclaved aerated concrete cakes being unloaded from a production autoclave

The behaviour of fly ash within autoclaved products is similar to the pozzolanic reactions that occur in conventional concrete and mortar. However, reaction rates are significantly higher and the crystallinity of the calcium silicate hydrate is increased. The reactions between siliceous pozzolanas and calcium hydroxide can be represented as follows:

$$xCH + yS = zH\, C_xS_yH_x + z. \quad (8.3)$$

The reaction products are calcium silicate hydrates (CSH phases). There is a reduction in the molar lime/silica (C/S) ratio of the CSH phases formed as autoclaving proceeds, with an associated increase in crystallinity. A CSH gel is considered to be the initial reaction product, which converts into semi-crystalline C–S–H (II) or C–S–H (I), which in turn crystallises into tobermorite ($C_5S_6H_5$).

Similarly, calcium hydroxide reacts rapidly with fly ash at typical autoclaving temperatures. However, since the pozzolana is an aluminosilicate, additional reactions occur. Calcined kaolinite ($Al_2O_32SiO_2$) may be used as an example of such a pozzolana, as

$$AS_2 + 3CH + zH\, CSH_{z-5} = C_2ASH_8 \text{ (gehlenite hydrate)}. \quad (8.4)$$

Amorphous alumino-silicate glass is the major constituent of fly ash. As is the case for a purely siliceous pozzolana, CSH phases form during autoclaving. However, phases from the garnet–hydrogarnet solid solutions series (C_3AS_3–C_3AH_6) also form. Aluminium substitution into the semi-crystalline CSH phases also increases the rate of crystallisation to tobermorite.

The formation of tobermorite and C–S–H (I) within the autoclaved matrix is associated with the build-up of compressive strength. These low-density CSH phases have a fine crystalline structure and numerous points of contact, which constitute bonding sites throughout the cured matrix. Hydrogarnets have dense octahedral crystals, which provide fewer points of contact and therefore contribute less to the strength of the autoclaved material.

Raw materials

Pulverised fuel ash

The chemical compositions and physical properties of some fly ash sources used to manufacture AAC in the UK are given in Table 8.1.

In general, 'run of station' ash is used for the manufacture of AAC. The fineness specifications of BS 3892 Part 1 or EN 450 are not relevant to autoclaved products. It is possible to use relatively coarse fly ash because the elevated temperatures and high alkalinity within the autoclaves ensure rapid dissolution of the aluminosilicate particles.

Table 8.1. Analyses of fly ash samples used to manufacture autoclaved aerated concrete

	Source A	Source B	Source C
Elemental analysis (%)			
SiO_2	41·36	45·74	55·62
Al_2O_3	23·67	25·45	24·20
Fe_2O_3	17·41	9·64	7·09
CaO	4·98	2·75	2·19
MgO	2·15	1·59	1·29
TiO	0·84	0·95	0·98
K_2O	3·07	2·81	2·52
Na_2O	1·01	1·18	1·27
MnO	0·15	0·08	0·05
P_2O_3	0·23	0·23	0·22
SO_3	2·34	2·22	0·84
C	2·79	7·36	3·73
Loss on ignition (%)	2·67	7·10	3·03
Median particle diameter (μm) (laser diffraction)	22·50	45·40	26·80
Relative density (g/cm^3)	2·36	2·02	2·15

The carbon content of fly ash, normally assessed by loss on ignition (LOI), has a significant effect on the casting and setting characteristics of AAC. High carbon ashes can have excessive water demands and give mixes with high W/S contents. A large number of carbonaceous particles also tends to destabilise the cellular structure of the cementitious mass. Both of these factors can lead to instability within cakes in the early stages of manufacture and collapse can occur. Wide variations in the carbon contents of fly ash can lead to significant process difficulties. However, these are controlled within the factory and the final product specification is maintained.

The composition of the fly ash also influences the hydrothermal reactions that occur during autoclaving. This affects the strength that can be achieved at a given density. In general, a 'good' fly ash in respect to its performance during autoclaving is characterised as having a high SiO_2 content, low Fe_2O_3 and low carbon (LOI). However, the performance of a fly ash sample cannot be predicted from its chemical analysis. Test mixes must be produced, autoclaved, and tested for strength and other relevant properties.

Cement

A range of Portland cements can be used for the manufacture of AAC. In the UK all cements comply with BS 12.[2] The choice between ordinary and rapid-hardening cement depends on the rate of setting required for the particular plant.

Quicklime (calcium oxide)

The choice of ground quicklime is an important consideration in the manufacture of AAC. Quicklime is produced by the calcination of either limestone or chalk under carefully controlled kiln conditions. The rate at which quicklime slakes to calcium hydroxide is termed its reactivity and is the most significant property of the raw material. The purity of the feed-stock, type of calcination kiln used, fineness of quicklime and amount of water added during grinding determine reactivity. Modern kilns generally produce medium- to high-reactivity quicklime of low variability.

Aluminium powder

Successful manufacture of AAC relies on the use of high-quality aluminium powder, often developed for particular formulations or combinations of raw materials. Finely divided aluminium flakes with modified surface characteristics must be used. During manufacture various organic additives are used and the aluminium powder is partially oxidised under controlled conditions. Various grades of aluminium powder are produced which liberate hydrogen at different rates.

Properties of autoclaved aerated concrete

General

AAC consists of air-filled cells, a result of hydrogen evolution, and a dense intercellular matrix. The matrix forms a network of narrow bridges surrounding the cells, accounting for only 20–30% of the volume of AAC, but responsible for its strength and, therefore, the load-bearing characteristics of the material (Fig. 8.7). The amount and nature of the reaction products within the matrix and its microporosity are important factors in determining the compressive strength of the bulk material.

Strength

The compressive strength of AAC determines the load-bearing characteristics of structures made with the material. There is a direct relationship between strength and density (Fig. 8.8).

The range of strengths illustrated is significant. At a given density, the compressive strength achieved depends on the choice of raw materials and the autoclaving conditions used. Experience of producing AAC has

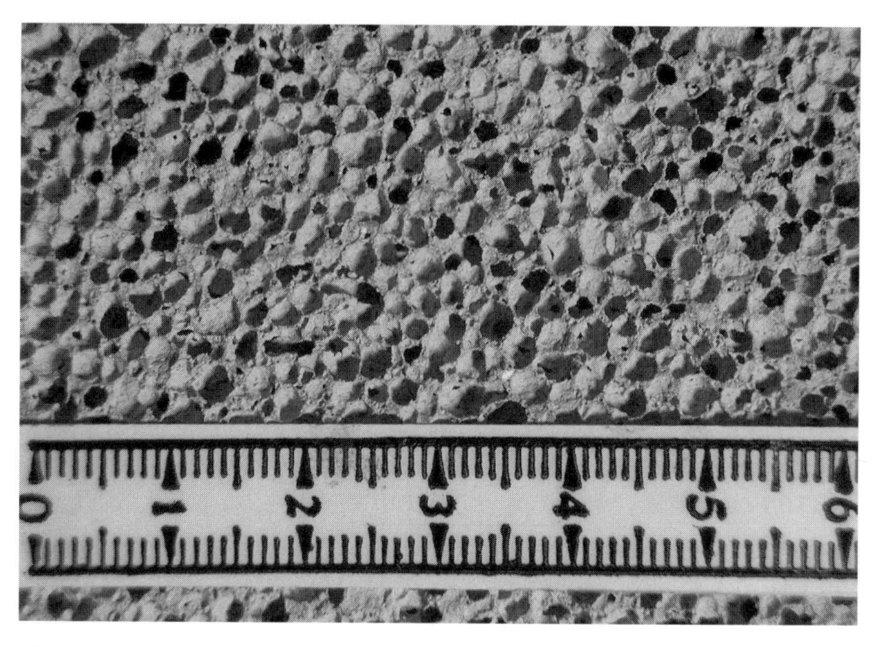

Fig. 8.7. Cellular structure of autoclaved aerated concrete

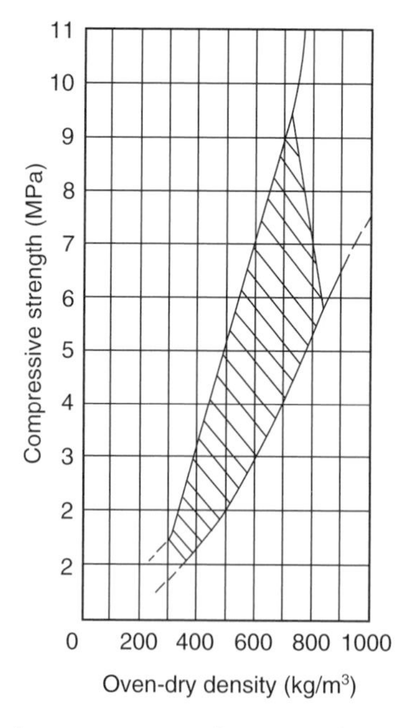

Fig. 8.8. Relationship between compressive strength and density for autoclaved aerated concrete (RILEM[3])

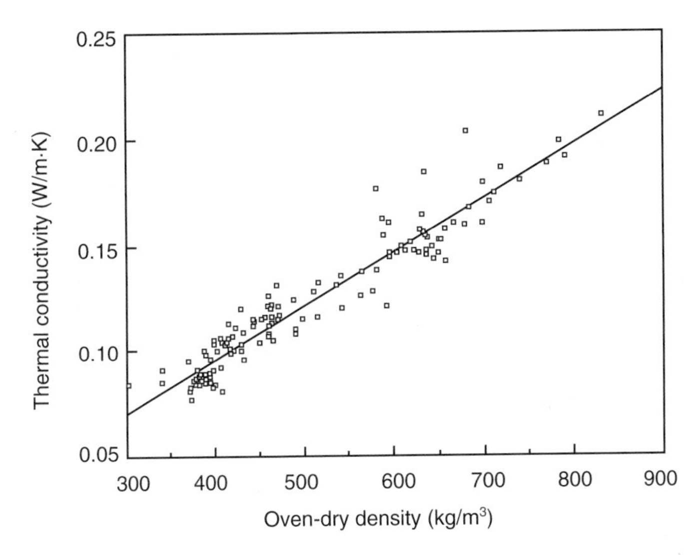

Fig. 8.9. Relationship between thermal conductivity and density for autoclaved aerated concrete (RILEM[3])

shown that the strength obtained for a specific autoclaving temperature and duration is affected by the siliceous raw material used. The particular fly ash source chosen often determines the ultimate compressive strength that may be attained.

Thermal conductivity

The thermal conductivity of AAC, at a particular moisture content, is directly related to density (Fig. 8.9). As a means of energy conservation, regulatory bodies in the industrialised countries have specified improved thermal insulation for new dwellings. This requires building products of lower thermal conductivity. For AAC this means lower density and the trend is likely to continue. However, the relevant standards for building materials, such as BS 6073, specify compressive strengths that must be achieved in order to maintain the load-bearing characteristics of structures. The selection of raw materials that maximise compressive strength at a particular density is therefore important.

Drying shrinkage

The drying shrinkage measured in AAC is influenced by the crystallinity of the CSH phases formed. The duration and pressure of autoclaving, raw material selection and formulation used therefore affect drying shrinkage. In the UK, BS 6073 Part 1 specifies a maximum value of 0·09% for AAC.

Durability

AAC has considerable resistance to frost damage. Ice crystals can form in cells, rather than within the matrix, and this reduces the disruptive pressures that build during freezing.

AAC made with fly ash has excellent resistance to attack by sulfate ions and two factors are important. First, crystals of reaction products can form harmlessly within the cells, similarly to the process that occurs during freezing. Secondly, the presence of significant amounts of aluminium from the fly ash leads to the formation of hydrogarnets during autoclaving, in addition to CSH phases. Hydrogarnet phases are resistant to sulfate attack and AAC specimens have been shown to survive class 4 conditions as defined by BRE Digest 363.[4]

Lightweight aggregate concrete blocks using furnace bottom ash

Market, applications and standards

During 1999 more than 2 million m^3 of lightweight aggregate (LWA) concrete blocks were sold in the UK for a variety of construction purposes in the density range of 1000–1500 kg/m^3. LWA blocks form cost-effective, general-purpose load-bearing or non-load-bearing masonry units with a proven track record for use above and below ground level. They can also be used as infill units in beam and block floor systems. Blocks can be produced in a standard face texture suitable for plaster and dry lining, a close-textured finish suitable for direct decoration or 'fair face' where no decoration is required (Fig. 8.10).

LWA blocks are available in solid, cellular and hollow formats depending on the design requirements. The common face size is 440 mm long by 215 mm high, with a thickness of 75–215 mm. Other face sizes are also available for different co-ordination options. Specially shaped blocks can also be produced to facilitate a variety of construction details.

Currently, BS 6073[1] is the relevant standard regulating the properties and testing of LWA blocks, and products may be assessed for the BSI kitemark against this standard. A harmonised European standard is in preparation, which covers a wider range of properties and will allow assessment for CE marking. This is likely to be published as BS EN 771-3.

Manufacture

The majority of LWA blocks today are manufactured in modern, highly automated and closely controlled factories (Fig. 8.11) using 'static' casting machines or mobile 'egg-laying' machines.

Fig. 8.10. Typical range of dense and lightweight concrete blocks produced in the UK

In a static plant, as shown schematically in Fig. 8.12, cement, fly ash, aggregates and water are usually batched into and mixed in a horizontal pan-type mixer to an almost earth-dry consistency which is then vibrocompacted into a mould on to a steel or wooden pallet. The blocks are immediately demoulded and transferred on the pallet to a storage area for curing. Commonly this curing area consists of enclosed chambers that are heated at temperatures up to 80°C, depending on the heat source, to accelerate the cement hydration and strength development to optimise the throughput of the plant. Blocks are then packaged into cubes (Fig. 8.13) of various configurations and further stored in a stockyard, if required, to achieve the final strength.

For an egg-laying machine the mix is transferred to the hopper of the mobile machine, which vibrocompacts the mix into moulds, momentarily set down on to a concrete base. Once the blocks are formed the machine lifts the mould immediately and moves along to the next position to repeat the process. The blocks are then left in place to cure until they are strong enough for handling and further storage. Egg-laying production takes place in large sheds or on outdoor concrete pads, depending on the local environmental conditions.

Fig. 8.11. Aerial view of a modern UK lightweight aggregate concrete block plant

In both static and egg-laying production, grading of raw materials, moisture content, cement content, vibration patterns and curing conditions are all key to quality, cost and rate of production.

Use of fly ash and furnace bottom ash in lightweight aggregate block production

Furnace bottom ash

Although many lightweight aggregates are suitable for use in LWA blocks, FBA has been used for many years and remains the most commonly used

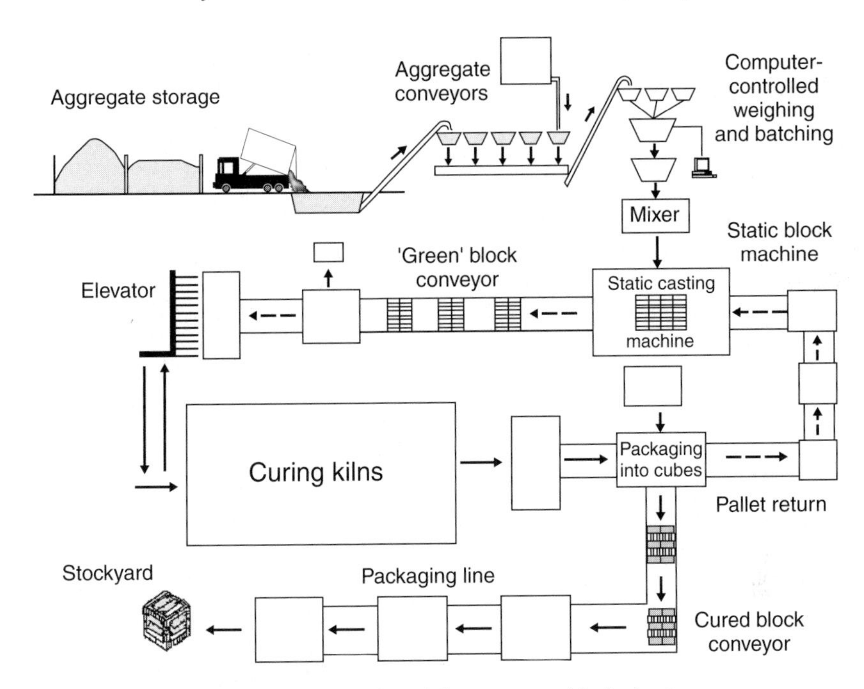

Fig. 8.12. Schematic layout of a lightweight aggregate block plant

material in the UK (Table 8.2), although the practice does not seem widespread in the rest of the world. The success of FBA in the UK may stem from the relatively limited range of chemical and physical properties of both FBA and fly ash.

FBA collected from the base of furnaces in the power station is usually water cooled (Fig. 8.14) and transferred to stock piles. The material is then crushed and screened in different size fractions either at the station or at the block manufacturer's site (see Chapter 1). Typical grades used in block production are 14-0, 14-5 and 5-0 mm, depending on the type of block being manufactured.

The density of FBA, in the range of 800–1100 kg/m^3 depending on the grade used, is ideally suited to achieve the typical LWA block density range and the relevant technical properties. For use in blocks, FBA is permitted by BS 6073. This gives limits on the properties specified in BS 3797[5] covering pumice, expanded clay, shale and slate, clinker and FBA, and those based on fly ash and blastfurnace slag. Table 8.3 gives the important values. FBA from most modern power stations is well within these limits.

Sintered fly ash aggregate

Pelletised/sintered fly ash lightweight aggregate (Lytag; see Chapter 7) can be used, thus increasing the utilisation of the major by-product from

Fig. 8.13. Automated plant forming the cubes of lightweight aggregate concrete blocks

Table 8.2. Utilisation of fly ash and FBA in the UK (1997 data)

Materials	Production (tonnes)	Utilisation (tonnes)	% in blocks
Furnace bottom ash	1,600,000	1,520,000	95
Fly ash	6,200,000	682,000	11
Totals	7,200,000	2,202,000	31

coal-burning stations. The aggregate produced, again easily in compliance with BS 3797, has a density of around 800–900 kg/m^3 and, therefore, can be used for blocks at the lower end of the typical density range.

The particle shape tends to be rounded which, when blended with other appropriate grades of aggregate, can be beneficial to 'workability' of the relatively dry block mix during manufacture. As production of sintered fly ash involves a secondary process, it is more expensive than FBA but can have the benefit of being more predictable and consistent in producing quality LWA blocks.

Fig. 8.14. Furnace bottom ash collection pits at a UK power station

Table 8.3. Limits on lightweight aggregates for masonry

Properties in BS 3797, lightweight aggregates for masonry and structural concrete	Maximum limit (%)
Loss on ignition (masonry units)	25
Reinforced and high-durability concrete	10
Sulfate content (expressed as SO_3)	1

Fly ash

BS 6073 permits the use of two types of PFA/fly ash to be used in LWA blocks:

- BS 3892 Part 1[6] PFA: This is the classified material primarily designed for use in concrete. It is classified to increase fineness and reactivity, and can reduce the water requirement of the concrete mix. These properties enhance the contribution of the fly ash to strength.
- BS 3892 Part 2[7] PFA: This is unclassified fly ash. It is lower cost as no processing is required. The economics make this option preferable in most cases.

In general, the fly ash used in the context of LWA blocks tends to be 'run of station' fly ash complying with BS 3892 Part 2. This is usually handled and transported to block plants in the dry powder form in tankers and silo-stored at the works. The function of the fly ash in this case is more as a fine filler to improve cohesion of the mix, which gives benefits during mould filling, compaction and demoulding, and in modifying the texture to improve the finish of paint quality blocks. However, depending on the source of fly ash, the other mix constituents and the efficiency of curing, there can be some contribution to strength from the pozzolanic reaction with cement. The inclusion of fly ash also contributes to block durability, particularly with regard to sulfate resistance.

Overall, the use of FBA and fly ash in LWA blocks has several benefits:

- sustainable conversion of by-products into a cost-effective building material;
- production of blocks with a wide range of properties and a flexible range of applications at a weight that is safe to handle manually;
- both FBA and fly ash are relatively inert materials, ensuring that the units produced are stable and durable using raw materials that can be stored and handled safely during manufacture;
- the environmental impact of block production is reduced, with the possibility of obtaining environmental credits, e.g. when using the BREEAM assessment method.

Fly ash in other precast products

Fly ash can be used either as part of the cementitious material or as an aggregate in numerous precast concrete products. These include paving units, kerbs, edgings and flags, concrete roof tiles, concrete pipes, lintels and fence posts. One example would be the UK section of the Channel Tunnel, where around 450,000 pre-cast reinforced concrete tunnel lining segments were produced. The units, cast with 30% fly ash of the cementitious component, gave a concrete mix of exceptionally low permeability and diffusibility.[8]

References

1. BS 6073-1. *Precast concrete masonry units. Specification for precast concrete masonry units*. BSI, London, 1981.
   BS 6073-2. *Precast concrete masonry units. Method for specifying precast concrete masonry units*. BSI, London, 1981.
2. BS 12. *Specification for Portland cement*. BSI, London, 1996. To be replaced on 1 April 2002 by: BS EN 197-1. *Part 1: Composition, specifications and conform ity criteria for common cements*. BSI, London, 2000.

3. RILEM. *Recommended practice, autoclaved aerated concrete, properties, testing and design*. E and FN Spon, London, 1992.
4. Sulfate and acid resistance of concrete in the ground. *BRE Digest 363*, BRE, Watford, July 1991.
5. BS 3797. *Specification for lightweight aggregates for masonry units and structural concrete.* BSI, London, 1990.
6. BS 3892. *Part 1: Specification for pulverised-fuel ash for use with Portland cement.* BSI, London, 1997. ISBN 0-580-26785-7.
7. BS 3892. *Part 2: Specification for pulverised-fuel ash for use as a type I addition.* BSI, London, 1996. ISBN 0-580-26444-0.
8. Eves RCW, Curtis DJ. Tunnel lining design and procurement. *Proceedings of the Institute of Civil Engineers, The channel Tunnel, Part 1: Tunnels,* 1997: 127–143.

Chapter 9

Other potential uses of fly ash

The use of fly ash with brick clays

There is nothing either new or startling about mixing ashes, resulting from the burning of coal, with clays for brick making. This has been done since the seventeenth century for the manufacture of stock bricks. The motives for doing so are to reduce the shrinkage of the clay and to make use of the heating value of the unburnt carbon in the ashes. The latter, derived from the minerals deposited with the vegetable matter that forms the coal, is very similar to that of clays. Anderson and Jackson[1] review the history of the use of fly ash in brick making and assess the results. They report that isolated trials with fly ash took place in the 1930s. Pilot-scale work was carried out in the early 1950s to establish the possibility of making bricks from a mixture consisting mainly of fly ash. Trials were also carried out with the minimum plastic clay plus fly ash to bond it and make it workable. This work was never carried through to full-scale production. For some years there has also been interest in calcined clay, i.e. clay lightly pre-fired to destroy its drying shrinkage and then ground and mixed with more clay from the brick-making body.

Outside that section of the brick industry that makes stock bricks, ashes are little used, although fired-clay grog is employed where necessary to control drying shrinkage. The range of composition of brick clay is very wide, as typically shown in Table 9.1. Since brick making is not a very demanding process in the materials sense, a variety of earthy materials can meet its requirements. The table also shows that fly ash is similar to brick clay in composition and, being a pre-ground, pre-calcined material, is a possible replacement for grog or calcine if its cost at the brickworks is not excessive. The amount and grade of fly ash added are dependent on the clay. Heavy clays require a greater quantity of fly ash and a coarser ash may be used.

Method of treatment to be followed

Before discussing the use of fly ash in brick making, some background information will be given on the brick properties and brick manufacture.

Table 9.1. Range of composition of brick clay and fly ash

Constituent	Brick clay (%)	Fly ash (%)
Silica (SiO_2)	80–43	58–38
Aluminium (Al_2O_3)	35–8	40–20
Iron (Fe_2O_3)	12–0	16–6
Lime (CaO)	26–0	10–2
Magnesium (MgO)	13–0	3.5–1
Potassium (K_2O)	6.5–1.5	5.5–2.0 (Potassium and Sodium combined)
Sodium (Na_2O)	1.5–0	
Sulfur (SO_3)	5.6–0	2.5–0.5

The use of fly ash as a minor constituent is compared to cases where fly ash is the major constituent and clay is a plasticiser. The possibilities of making structural ceramics other than bricks will be considered briefly, as will the possible utilisation of cenospheres.

Properties required of bricks

General

Fly ash can be used successfully in brick making only if the resulting bricks have the required properties. It is necessary to distinguish between intrinsic properties of the bricks, such as density and soluble salt content, and performance properties, such as weather resistance. The latter depend not only on the bricks, but also on the applications.

Methods of testing are prescribed in BS 3921,[2] in which the variations in the conditions that bricks have to withstand are described:

- 'ordinary quality': for normal applications in internal walls
- 'normal exposure': in external walls between the damp-proof course and the roof
- 'severe exposure': in parapets, retaining walls and below the damp course where the bricks may become and remain saturated and may be frozen in that condition.

This broad distinction is meaningful over the greater part of the British Isles, although there are exceptions, e.g. in wet mountainous areas a brick of 'special quality' may be required in external walls.

Physical and chemical properties

The physical properties of bricks that are most commonly measured are compressive strength and water absorption. Compressive strength is required primarily as a basis for calculating permissible pressures on load-bearing brickwork.[3] Water absorption, when measured by the vacuum method, is expressed as a percentage by volume or weight. The only direct application of the water absorption test is the identification of bricks that have so low an absorption ($<45\%$ by weight by the vacuum method) that they can be used in damp-proof courses.

The chemical composition of bricks is not critical, but it is important that the proportion of water-soluble sulfates, especially magnesium, sodium and potassium sulfates, should not be too high when the bricks are to be used in very wet conditions. Sulfates can lead to efflorescence and sulfate content should be tested for. The danger is that a chemical reaction between the sulfate and tricalcium aluminate in the set cement, or hydraulic lime, may cause the mortar to expand and, ultimately, to soften. BS 3921[2] imposes limits on soluble salt content for bricks of special quality suitable for these conditions. No such limits are imposed for bricks of ordinary quality, but even in normal exposure bricks containing significant quantities of the more soluble sulfates are to be avoided. Fly ash often contains calcium sulfate, but has not been found to contribute dangerous quantities of sulfates to bricks in which it has been used.

Performance characteristics

Bricks are the chief constituents of walls forming part of structures, which provide permanent shelter. These walls are therefore required:

- to be sufficiently strong and stable
- to resist rain penetration
- to provide sufficient thermal insulation
- to be dimensionally stable, ensuring freedom from cracking and distortion
- to be durable.

Most of these performance characteristics are only partly dependent on the intrinsic properties of the bricks. Thus, the strength and stability of walls depend primarily on their being of sufficient thickness in relation to their height and being supported laterally by cross-walls. The intrinsic strength of bricks is seldom fully used, although there is increasing interest in load-bearing brickwork.

Resistance to rain penetration is sometimes thought, wrongly, to depend on the bricks being of low absorption. In fact, it is now usual, except in sheltered positions, to rely on cavity construction to prevent rain penetration. Even if a solid wall is built, it is by no means certain that a dense brick will give a drier wall than a porous one. Indeed, a permeable brick, such as many of those containing fly ash, has a certain capacity for water, which may be beneficial in that it may hold any absorbed rainwater until it can evaporate later.

Thermal insulation also largely depends on the choice of wall construction, on the presence and thickness of the cavity, and on whether or not insulation is used. For example, with average bricks, a 300 mm cavity wall has a thermal resistance one-third greater than that of a 230 mm solid wall, although the thickness of brick material in it is the same. Thermal resistance also varies significantly with the porosity of the bricks, with more porous bricks showing higher thermal resistance.

The dimensional stability of brickwork depends on both the bricks and the mortar. In addition to thermal expansion and contraction, which are normally well understood and allowed for, it is necessary to consider moisture expansion of bricks and expansion of mortar due to sulfate attack. It has long been known that all masonry materials undergo an expansion on wetting and a shrinkage on drying, and that these movements are generally much smaller with fired-clay products than with mortars, most natural stones and concretes. In addition, fired-clay products have a somewhat larger irreversible expansion owing to the absorption of moisture vapour from the air. This expansion is relatively rapid in the first few days after the bricks leave the kiln. It proceeds at a decreasing rate for months or years; it is also independent of the small reversible wetting and drying movement and is not extinguished by wetting the bricks in water. The exact amount of the expansion varies with the nature of the clay from which the bricks are made and with the firing temperature. No precise information on the subject was available in the early 1950s when, as noted above, most of the pilot-scale work on fly ash bricks was done.

Durability, in the UK climate, is primarily a matter of frost resistance. Frost resistance depends at least as much on the conditions of freezing, and especially on how wet the bricks are when frozen, as it does on the properties of the bricks. It is often assumed that high compressive strengths and low water absorption are signs of good durability. There is some truth in this idea, but there are many exceptions. Many bricks of high water absorption and comparatively low strength are durable. In addition, fairly dense and strong bricks are liable to decay in severe exposure. It is necessary to stress this point because bricks containing a substantial proportion of fly ash are frequently weaker and more porous than bricks made from the same clay without fly ash. However, this does not usually make the brick containing fly ash less resistant to frost; indeed, the reverse is often true.

Requirements of manufacturing processes

Cost limitations of brick making

The selling price of common bricks and good-quality facing bricks is relatively low. Bricks are probably the cheapest manufactured articles in existence. Manufacturing on a large scale in modern and very efficient plants by simple and largely automatic processes keeps the costs low. Clay can be dug in quantity, often from a pit beside the factory for minimal cost, with between 3 and 4 tonnes being required for 1000 bricks. The labour requirement, in an up-to-date plant, is about 3 man-hours per 1000 bricks. These figures will indicate that, if fly ash is to be used in brick making, the brickworks must be near to suitable sources to avoid high transport costs. The scale of manufacture necessary to achieve the standard of efficiency indicated above is shown by the size of plants and the high capital costs. Few plants have a capacity of fewer than 300,000 bricks per week. Facing bricks and other high-quality bricks command higher prices and they are considered essential to the economic success of a modern plant.

Processes of manufacture

The first stage in the brick-making process consists of grinding, screening and tempering (i.e. adjusting the water content) of the clay. Next follows the forming of the bricks usually by extruding a ribbon of plastic clay and wire-cutting it. There are other methods in use in the UK, but less so in most other countries. Where the clay used is a mudstone or shale, which is dry enough to be ground to a granular form, the bricks may be semi-dry pressed (the well-known Friction brick is the typical representative of this process) or stiff-plastic pressed. The most typical stiff-plastic-pressed bricks are common bricks made from colliery shale and some smooth red bricks made from shales from the coal measures. Where the clay is a soft and silty material of recent geological age, it may be machine moulded, as in making London stock bricks. When the bricks have been formed, they must be dried and fired. Wire-cut bricks are usually, and moulded bricks must be, dried in a separate dryer. Semi-dry-pressed bricks, stiff-plastic-pressed bricks and some wire-cut bricks made by stiff extrusion are stiff enough to be stacked in the kiln without deformation. They can then be dried and fired without being rehandled in between. Nearly all of the most modern plants use car tunnel kilns in which the bricks are passed through a stationary fire on cars with firebrick decks. Most of the older plants use one of the many variants of the continuous kiln invented by Hoffmann in 1862. The fire travels round a closed circuit, with fired bricks being removed and green bricks set at the point farthest from the fire. Kilns may be fired with coal, oil or gas and it is usually advantageous if some fuel is contained in the bricks. One of the reasons for using fly ash in brick making is that any carbon that remains unburnt can be oxidised in the time available in a brick kiln, and its heat utilised.

Manufacture of fly ash bricks

General

It may be possible to use one particular ash with a particular clay, and the ash may show advantages in both manufacture and the properties of the bricks produced. Yet, with different clay, even from the same works, the introduction of the ash may lead to increased losses and lower quality. It is always necessary, therefore, to consider specific cases.

Two general situations may be distinguished:

- those in which fly ash is the minor constituent, being added to the clay to improve its working properties and for its fuel value
- those in which the fly ash is the principal constituent, the clay being essentially a bonding agent.

There is no theoretical reason why there should not also be 50/50 mixtures of the two materials. Some have been used, although in practice it seems that the clays ordinarily found to be suitable for brick making are of such a size-grading and plasticity that they do not readily accept more than 10–30% by volume of fly ash without some deterioration in their properties. Hence, additions of this order are common when the use of fly ash is taken up by existing brickworks. At the other end of the scale, when the aim is to use the maximum amount of fly ash, it is possible to find bond clays which will enable the ash to be worked with the addition of only 15–20% of clay. This was the type of mixture considered when proposals were studied for constructing brickworks near power stations as a means of fly ash utilisation. Here, it was the clay that would have had to bear the cost of transport and there was interest, therefore, in keeping the proportion of clay to a minimum. Although hundreds of bricks of very satisfactory quality were made in the large pilot-scale trials, no works has been established in Britain to use the process on the full scale. The problems of full-scale production of bricks consisting mainly of fly ash remain to be worked out. Anderson and Jackson[1] suggest that high ash bodies are possible (>65% fly ash) and with lower levels of fly ash (25–35%) a comprehensive range of clays could be used if attention were paid to the mix design.

Addition of fly ash to brick-making clay

Reasons for use of fly ash

The results obtained by the addition of fly ash to brick clay depend largely on the grading or particle size distribution of the materials. Brick clays in general are far from consisting entirely of 'clay' in the sense of material of a particle size below 2 μm equivalent diameter. Some such as the Wealden series, found by Watts[4] to carry fly ash satisfactorily, may contain about 60% of the clay fraction, but others such as the Keuper marl have been found to contain only 12–35%.

Fly ash is a coarser material than brick clay. It therefore follows that the maximum scope for the use of fly ash is with those clays having a high content of material below 2 μm equivalent diameter. Each case will have to be decided by tests on the materials to be used. Where a suitable combination of brick clay and fly ash is found, one or more of the following advantages will accrue:

- Easier feed to the grinding mill: this usually only applies in the case of very wet or sticky clays.
- Less power required for extrusion or pressing: Watts[4] carried out a works trial with Weald clay in which some 200,000 bricks were made. He found the average power produced by the engine driving the brick machine to be 4–3% higher when running without fly ash than when an addition of 11–5% by weight (20% by volume) was made.
- Easier drying.
- Savings in fuel: in the trials conducted by Watts[4] there was a saving of 45 kg of coal per 1000 bricks fired, which represents a reduction of one-third of the coal consumption compared with when no fly ash was used.
- Easier firing: in the same trials, it was found that the fire was more easily controlled and travelled more rapidly. This was supported by the results of another trial[5] using a 60 : 40 mixture of Keuper Marl and fly ash. In the case of shales with a high carbon content, which are used for brick making in some places, fly ash is sometimes used as a diluent to reduce the carbon content to acceptable limits.

Limitations of works trials

In many works trials with fly ash the number of bricks made has usually been at most a few thousand, enough to set about half a chamber of an average continuous kiln. Consequently, it has not been practicable to vary the firing temperature from that normally used without fly ash. This, however, may not be the most suitable temperature for the clay/fly ash mixture and such a trial may therefore give an unduly unfavourable impression of the quality of bricks that can be made. This is confirmed by evidence given below, where the properties of laboratory-fired bricks are compared with those fired at the works.

Handling and treatment of fly ash

The method of handling to be adopted will depend to some extent on the conditions of supply and the equipment of the works. Three types are possible:

- Delivery to the clay pit: Damp fly ash is tipped into the clay pit beside the clay haulage and the required amount placed in each tub before it is filled with clay. The contents of the tubs are then tipped into the grinding mill in the usual way. This method gives the maximum period of mixing and may be satisfactory so long as the tipped fly ash does not dry out. Suitable precautions against this should be taken, as otherwise there may be some dust nuisance in dry weather.
- Delivery to the preparation machinery: Damp fly ash is fed in a regular stream into the train of preparation machinery at a convenient point where adequate control can be maintained. For example, it can be fed to the mixer with the ground clay, particularly if the mixer is or can be enclosed. Before adopting this procedure the adequacy of the mixing available between the point of supply and the brick machine should be checked.
- Deliveries of dry fly ash: The cost of the handling plant tends to limit this to works of high output.

Fly ash bricks with a clay bond

Scope of pilot-scale experiments

Between 1952 and 1955, extensive tests were conducted on two particular mixtures of fly ash and clay:

- fly ash from Hams Hall, near Birmingham, with an Etruria Marl from Wilnecote
- fly ash from Rye House, Hoddesdon, with London Clay from Nazeing.

These two case studies covered a wide range of conditions. The Hams Hall fly ash was available in the dry state and the problems of dry mixing it with ground clay were studied. At Rye House, fly ash was available only as slurry and involved the quite different problems of dealing with an ash of high water content. Hundreds of bricks were made successfully from each combination of materials, generally in the proportion of 85 : 15 fly ash : clay by volume (80 : 20 by weight). In the absence of experimental brickworks, no observations were possible on the drying and firing of the bricks under practical conditions. However, the bricks were fired in the laboratory to different temperatures for the studies of their properties, as reported below.

Manufacture based on dry fly ash

Several attempts were made to produce semi dry-pressed bricks from a mixture of fly ash with ground clay. Some bricks of tolerable appearance were produced, but their quality as shown by exposure tests was so poor that it is unnecessary to discuss them further (see the photographs of exposure piers in Figs 9.1 and 9.2). The wire-cut process achieved the best results.

With Hams Hall fly ash, it was found that the dry ash mixed with dry powdered clay, if wetted down immediately before extrusion, had remarkable working properties. It tended to stiffen up after extrusion, so that the green bricks were resistant to damage in handling and were very easy to dry, with a linear drying shrinkage of only about 2%. It was possible to put freshly made bricks into an oven at 70°C and to dry them in about 6 h, a treatment that few ordinary clay bricks would withstand. It is probable that the early stiffening, and the insensitivity to drastic conditions of drying, was due to the early removal of some of the mixing water. This was by absorption on the large surface area of glassy particles in the fly ash. In addition, water was lost by combination with the small amount of anhydrous calcium sulfate in the fly ash. The same mixture of fly ash and clay did not possess the same advantages if it had been kept wet overnight before extrusion. The bricks made from the mixture so treated were much more sensitive to drying conditions.

A corollary of the unusual stiffening behaviour of the newly wetted mixture was that care had to be taken to prevent it from setting in the brick machine. It would normally be necessary to run the machine empty at the end of a day's work, and possibly before a lunch break, to avoid difficulties in restarting.

Manufacture based on wet fly ash

These experiments were conducted using fly ash from Rye House power station, where the ash was discharged wet to a lagoon, and could not be readily extracted from the system in the dry state. The preparation of a mixture of fly ash and clay suitable for brick making was more troublesome and likely to involve higher costs for plant than at a power station where dry ash is available. The basic difficulty with fly ash from a lagoon is the high water content. It remains even when the fly ash has been allowed to settle in a tank and as much water as possible has drained away. Under such conditions, a water content of at least 55%, calculated on the wet weight, is likely. Further reduction in the water content is essential, and vacuum filtration is capable of bringing the water content down to about 40% rapidly. If the bond clay were then introduced as a dry powder, the result would be a mixture that might be suitable for machine moulding. It is likely not to be sufficiently stiff for the wire-cut process unless more water is removed. As has been emphasised previously, any instance where the idea of making bricks from fly ash with a clay bond is contemplated must be considered on its merits, after trials with the proposed materials.

It will be seen, however, that the possibilities are more restricted if the fly ash has to be taken from a lagoon, and the prospects of an economic and successful process are much reduced.

Firing of fly ash bricks

It has been pointed out that even a modest addition of fly ash effects a worthwhile saving in the fuel consumption necessary for firing bricks. It follows that some attention must be paid to the carbon content of any fly ash that might be used for a brick containing 70% or 80% fly ash. This is a problem which is already familiar to makers of colliery shale common bricks, since many colliery shales contain more coal than is really required to fire the bricks. The result is that the air supply to the kiln has to be restricted to prevent overfiring. The bricks tend, consequently, to have black cores owing to insufficient oxidation. The same situation could arise with fly ash bricks if an ash of high carbon content was used, and this would make it impossible to produce good-quality facing bricks. At base-load power stations, it is perfectly possible for the carbon content to remain below 4% for long periods. Provision should be made for the prompt detection, and exclusion from supplies going to the brickworks, of any fly ash of higher carbon content that may be produced.

Special bonding materials

The materials considered so far as bonds for fly ash have all been common clays, such as may be considered for brick making in the absence of fly ash. If the considerations mentioned above are borne in mind, it would be difficult to justify the use of anything more expensive. It is necessary, however, to mention one or two alternatives that have been proposed, and have been the subject of some experiment.

In the late 1940s, an attempt was made to produce floor tiles from fly ash by burning out the residual carbon (thereby losing its potential fuel value), bonding with sodium silicate, pressing and firing. Some small, laboratory-made tiles of indifferent quality were made, but an attempt to make tiles of normal size on a press of the type used for clay tiles was a complete failure. It is possible that the use of a much higher pressure on a different type of press would have given better consolidation.

Consideration had to be given to the subject matter of the Corson patent.[6] This proposed making bricks from fly ash with a variable proportion of furnace-bottom ash and with 1–2% of bentonite as a bond. Bentonite is a clay mineral of exceptionally fine particle size and high plasticity which finds application in drilling muds and as a bond in foundry sands, to quote only two examples. These make it probable that it may be effective even in the very small proportions quoted. Fuller's earth is mineralogically very similar to bentonite, and four different grades of fuller's earth were tried on a small experimental scale, with bricks being made by hand and fired in the laboratory. It was found that 1% and 2% of fuller's earth were not enough to bond fly ash effectively. With 5%, the absolute maximum that could be considered on grounds of cost, some rather weak and friable bricks were made which nevertheless were more weather resistant than expected. It was clear, however, that 15% of plastic common clay, such as London clay, would be much cheaper than 15% of Fuller's earth and would make a better brick. A more promising approach appeared to be the substitution of sodium silicate for the fuller's earth or bentonite.[7]

Apart from the special bonding materials discussed above, there are two approaches to the production of fly ash bricks that appeared to be promising, but have led to disappointing results. There are theoretical reasons why this should be so and again a warning may save wasted effort. These two approaches are semi-dry pressing and the total omission of the clay bond.

The making of bricks by semi-dry pressing has one main attraction, namely that it practically eliminates drying shrinkage and hence the difficulties of cracking and warping that can arise in drying plastic articles made from clays of high shrinkage. In spite of this, plastic processes and especially the wire-cut process have maintained their position nearly everywhere, the semi-dry-pressed Fletton brick being an exception for which there are special reasons. It is very difficult, in the semi-dry process, to secure effective bonding of the grains of clay. They tend to stick where they touch, leaving pore spaces in between which can be easily opened up by frost if the bricks are used in conditions of severe exposure. The imperfect contact between the grains means that the firing process is less effective in developing the all-important partial fusion of different mineral species. When the same material is made using the plastic process, the grains are in more intimate contact. Normally, a semi-dry brick needs a higher firing temperature than a plastic-made brick to mature it, and more accurate temperature control is necessary to avoid overfiring. Since bodies containing high amounts of fly ash have very small drying shrinkage, there is no real reason to depart from the plastic process.

When, in addition to adopting the semi-dry process, the bond clay is omitted and an attempt is made to produce bricks from moist fly ash alone, the difficulties are multiplied. The very fine powder is difficult to consolidate by pressing without developing pressure cracks due to entrapped air, although bricks can be made, at least from some samples of fly ash, by double pressing. Even with evacuation of the press die, difficulties have been experienced in eliminating the considerable volume of air entrained by the damp fly ash. The addition of grog does not fully overcome the difficulties. Methods of doing so have been suggested in work carried out in the laboratories of the Central Electricity Generating Board.[8,9]

Properties of fly ash bricks

Properties considered

When the original exploratory work on fly ash bricks was done,[4,5] the conventional tests for water absorption and compressive strength were used as a rough method of assessment, in spite of their known shortcomings. Analyses of soluble salts were conducted on a number of samples in which London clay was used, because soluble salts are frequently a problem with clays of that formation, but not on all samples. These analyses will be considered separately from the physical properties. In addition, bricks from the various test batches were exposed to the weather under standard severe conditions, and useful information has accrued from these exposure tests.

The form of exposure test generally used was the 'tray' test, in which representative specimens, usually two, were placed on edge in metal trays 75 mm deep and left outside. During the winter, the trays normally contained water, so that the bricks were saturated when frozen. In addition, similar specimens were exposed on a level concrete slab outside the trays. The tray test is very severe and bricks decay in it in roughly one-third the time needed for similar effects to occur in the brick on the edge coping of a parapet. Exposure outside the trays is not as severe as in the trays. The condition of the bricks in exposure tests is recorded in Tables 9.2 and 9.3 as good (G), moderate (M) or bad (B), according to their condition after 7 years' exposure. The mark G means no significant decay in 7 years, M means decay to a depth not exceeding about 25 mm, and B means decay that is more extensive.

In addition, a few brick piers were built. These were built and capped in such a way that rainwater drained from the capping into the middle of the pier, which was filled with sand. The conditions were classified as severe exposure, comparable with an earth-retaining wall (Figs 9.1 and 9.2). Fig. 9.1 shows a wire-cut rustic facing brick and Fig. 9.2 a semi-dry-pressed common brick, both containing 85% by volume of fly ash, after 12 years' exposure. The condition of these piers emphasises the point that it is difficult to make durable bricks from fly ash by semi-dry pressing.

Properties of bricks with moderate additions of fly ash

Reasonable comparisons are available for the effect of additions of fly ash on the properties of two types of clay, each represented by two examples. The types are:

- coal measure shales (Carboniferous)
- Keuper marl (Triassic).

Fig. 9.1. Rustic wire-cut facing bricks, containing 85% by volume fly ash, after 12 years' exposure as a hollow pier

Table 9.2 shows that the effect of fly ash on the properties of the bricks varies from clay to clay. With some clays, the porosity and strength are hardly affected at all (within the limits of variation to be expected in works tests), whereas with others the porosity is increased although the strength is not necessarily decreased. Even where the porosity is increased and the strength decreased, the durability of the bricks in severe exposure is not affected adversely; indeed, it seems to be marginally better. (The number of bricks used in the exposure test is small, so it is difficult to be sure of small differences in performance.) The marked improvement in the properties of the Keuper marl brick fired in the laboratory, compared with the same mixture fired at the works, is to be noted. The temperature of the laboratory firing was adjusted to the needs of the fly ash mixture and was

Fig. 9.2. Semi-dry-pressed common bricks, containing 85% by volume fly ash, after 12 years' exposure as a hollow pier.

about 50°C higher than the temperature used in the kiln at the works for firing the clay without fly ash.

Properties of bricks with a high proportion of fly ash

Information about the properties of bricks made mainly from fly ash with a clay bond rests chiefly on the results of pilot-scale tests in which the bricks were fired under laboratory control. One or two trials with a high proportion of fly ash have been conducted under works conditions. There is no point in attempting a comparison with the properties of the bond clay without fly ash, since the clays selected for this purpose are not

Table 9.2. Effect of moderate proportions of fly ash on the properties of bricks

Source and brick type	Bricks containing no fly ash				Bricks containing fly ash				
	Water absorption (% by weight)	Compressive strength (MPa)	Condition after 7 years		Fly ash and Clay (% by volume)	Water absorption (% by volume)	Compressive strength (MPa)	Condition after 7 years	
			In tray	Outside				In tray	Outside
Carboniferous clays									
Hamstead SP	18·3	21·9	B, B	M, B	40/60	18·2	23·5	M, B	G, M
Tamworth	12·2	31·9	G, G	G, G	40/60	18·6	36·8	G, B	G, G
Keuper marl									
Knowle WC	22·7	25·3	B, B	B, B	50/50	19·7	25·6	B, B	G, B
Bickenhill WC	20·8	32·9	B, B	B, B	39/61	27·0	18·8	B, B	B, B
	21·5	28·4	B, B	B, B	50/50	27·2	16·9	B, B	M, B
Bickenhill WC, Lab. fired					39/61	21·6	30·8	B	G

SP: stiff-plastic pressed; WC: wire cut; G: good (no significant decay); M: moderate (decay not exceeding 25 mm); B: bad (extensive decay).

Table 9.3. Properties of bricks with a high proportion of fly ash

Materials	Fly ash and clay (% by volume)	Method of making	Firing	Water absorption (% by weight)	Compressive strength (MPa)	Condition after 7 years	
						In tray	Outside
Hams Hall fly ash and Etruria marl	85/15	M	Works 'soft'	29·2	17·3	G, G	G, G
	85/15	M	Lab. 1080°C	25·5	25·9	G, G	G, G
	85/15	HM	Lab. 1080°C	26·8	23·3	G, G	G, G
	85/15	HM	Lab. 1070°C	30·2	18·5	G, G	G, G
	90/10	HM	Lab. 1070°C	31·6	15·3	G, G	G, G
	95/5	HM	Lab. 1070°C	35·5	–	G	M
	85/15	WC	Lab. 1080°C	22·1	31·4	M, M	G, G
	85/15	SD	Lab. 1000°C	29·0	10·7	B, B	G, B
Rye House fly ash and London clay	85/15	HM	Lab. 1080°C	32·7	15·4	G, M	G, G
	85/15	WC	Lab. 1080°C	30·7	22·8	M, M	G, M
Portobello fly ash and coal measure shale	75/25	WC	Works	15·0	39·7	G	G
Fly ash and Brickearth made stock fashion	75/25	M	Works	31·6	19·8	G, M	G, G
Rye House fly ash and fuller's earth (four different grades)	95/5	HM	Lab. 1080°C	48·4	–	G	G
	95/5	HM	Lab. 1080°C	46·9	–	G	M
	95/5	HM	Lab. 1080°C	47·3	–	G	G
	95/5	HM	Lab. 1080°C	46·6	–	B	M

M: machine moulded; HM: hand made; WC: wire cut; SD: semi-dry pressed; G: good (no significant decay); M: moderate (decay not exceeding 6 mm); B: bad (extensive decay).

necessarily suitable for brick making on their own. No bricks were made from the clays alone. Representative test results on two examples are given in Table 9.2.

Bricks with a high ash content have properties very similar to those of many other facing and common bricks in general use, although their water absorption levels are generally very high. While an engineer accustomed to the dense bricks from the coal measures may regard these high water absorption figures as unacceptable, Table 9.3 contains evidence that this attitude would not be justified. The excellent durability in the very severe conditions of the tray exposure test of nearly all the specimens tested would compare favourably with any random selection of ordinary clay bricks. It is not easy to find satisfactory reasons why these fly ash bricks are as good as they are. A possible explanation may be found in their generally uniform pore sizes, which at least theory supports. Frost action takes place in a structure in which ice lenses in coarse pores can be fed by water in adjacent very fine pores. It should not be overlooked that the high porosity of these bricks of high fly ash content will have a very favourable effect upon their thermal insulation properties.

Soluble salts content of bricks of high fly ash content

As mentioned above, London clay has a bad reputation as a source of efflorescence and other problems due to soluble salts in bricks. The possibility of excessive amounts of soluble salts being found in London Clay/fly ash mixtures was therefore studied carefully. Sample bricks were made from fly ash and clay taken from different depths in a number of different boreholes. They were fired to 960°C, 1020°C and 1080°C, i.e. over the range which could be practically described as moderately well fired to hard fired, and were analysed. Similar trends were shown by all the batches tested, and are illustrated by the results in Table 9.4.

Table 9.4. Analyses of soluble salts of fly ash and London clay bricks

	Temperature (°C)		
Bricks fired to:	960	1020	1080
Total soluble salts composition	2·58	2·31	0·99
Silicate (SiO_2^{2+})	0·04	0·03	0·01
Sulfate (SO_3^{2+})	1·50	1·42	0·65
Calcium (Ca^{2+})	0·71	0·59	0·28
Magnesium (Mg^{2+})	0·01	<0·01	<0·01
Sodium (Na^+)	0·03	<0·01	0·02
Potassium (K^+)	0·03	<0·01	0·03

Since the soluble salts consist mainly of calcium sulfate, and the amount is well within the limits found with other bricks in common use, e.g. Flettons, they provide no grounds for alarm. At the same time, with this clay at least, a uniformly high firing temperature should clearly be the objective.

Hollow blocks

With the exception of one case near Birkenhead,[10] little development work has been done on the manufacture of structural ceramic products other than standard solid bricks from fly ash mixtures, but the general possibilities can be indicated. The plasticity that the raw material must possess for the manufacture of different products increases in the order: solid bricks $<$ perforated bricks $<$ hollow blocks (and/or roofing tiles) (Fig. 9.3). It is not very useful to attempt to quote actual figures for the plasticity necessary for each type of product. The ranges for plasticity are wide and the plasticity at which a given type of product can be made successfully depends on the machinery used and the skill with which it is operated. However, it will be useful in any programme of practical trials to determine the plasticity index of any mixture on which extrusion trials are conducted. In this way, a useful collection of basic data can be built up as the programme proceeds.

In the mixture of moderate fly ash content, it is unlikely that the presence of fly ash will prevent a manufacturer from producing perforated

Fig. 9.3. Some products that have been made using fly ash

bricks owing to lack of plasticity. Fly ash will be used if the clay is plastic or even sticky and the mixture chosen for solid bricks may well be usable without modification for perforated bricks.

In high fly ash mixtures, much will depend on the bond clay available in any particular instance. It may be necessary to increase the proportion of bond clay, from one-fifth to perhaps one-third, to obtain the plasticity necessary for hollow-ware. The possibilities still need to be worked out. The manufacture of hollow blocks of some kind may be a useful way of keeping control of the firing process in case the carbon content of the fly ash tends to rise. A kiln set with hollow blocks necessarily contains a smaller weight of ware per volume of kiln space than one set with solid bricks. Therefore, the use of a slightly increased proportion of bond clay may enable a slightly higher carbon content in the fly ash to be tolerated.

Fly ash in asphalt

Hot rolled asphalt (HRA) is widely used for the surfacing of motorways and heavily trafficked roads. It is produced by heating a mixture of mineral aggregates, filler and bitumen to relatively high temperatures. The material is transported, laid and compacted in a short period to avoid loss of temperature. HRA is produced at 1600°C, placed and compacted at not less than 1250°C, to maintain workability. Proportions of coarse aggregate (34%), sand (56%) and filler (10%) required to produce specified size distributions are given in BS 594 Part 1, 1992. In general, the filler used is limestone dust.

A detailed laboratory study followed by a site trial on a heavily trafficked road was undertaken using HRA with fly ash replacing the limestone dust filler. Mixes were prepared at various mixing and laying temperatures. These temperatures were recorded and subsequently the paved area was examined for rutting, surface texture and density.[11]

The predominantly spherical particles of fly ash improved the packing properties of HRA, and since they tend to occupy more bulk volume per unit weight, a lower bulk density was achieved. The fly ash HRA had a far higher workability index than conventional asphalt. This suggests that the fly ash could be mixed and compacted at temperatures as low as 110°C and 850°C, respectively without impairing its engineering performance and other properties.

Savings in energy input when replacing limestone dust with fly ash were considerable, and this classifies the product as a low-energy material. The optimum bitumen content for engineering properties was found to be lower. However, there was some concern about reducing the bitumen content for durability reasons, especially the oxidation of bitumen. Replacing the limestone dust with fly ash did not affect the optimum bitumen content of the HRA.

Other ceramic products

Cenospheres

When fly ash is discharged into a lagoon, a small proportion of the particles separates from the rest of the material and rises to the surface. These are the cenospheres, which are hollow particles made of expanded silicate glass. They have a non-porous shell thickness of about 10% of their radius. The mean diameter is 100 μm, the range (by weight) being 5% below 50 μm and 20% above 125 μm. Typically, 1–2% of the fly ash produced from the combustion of coal in power stations is formed as cenospheres. Their main characteristics are:

- hollow spheres with spherical morphology
- particle sizes ranging from submicrometre to millimetres
- ultralow density
- low thermal conductivity
- high particle strength
- resistance to acids
- low water absorption.

Applications

The main application is that of an inert filler. With a density lower than water (typically 0·7), cenospheres provide up to four times the bulking capacity of normal-weight fillers. The microspherical shape dramatically improves the rheology of fillers, whether in wet or in dry applications. It is an extremely stable material. It does not absorb water and is resistant to most acids. As it is a refractory material, it can resist high temperatures.

Cenospheres can be used in plastics, glass-reinforced plastics, lightweight panels, refractory tiles and almost anywhere that traditional fillers can be used. Because of their flexibility, they are used in many high-technology and traditional industries. Aerospace, hovercraft, carpet backing, window-glazing putty, concrete repair materials, horticulture, and off-shore oil and gas production industries all use cenospheres. They are also used as an aggregate for insulating blocks and an additive to fire clay to aerate it. When the clay is heated to about 1400°C the cenospheres collapse, leaving a cellular structure with improved insulating qualities. Both of these applications have been exploited in Europe.

In the aerospace industry, cenospheres have been used to manufacture lightweight propeller blades. The cenospheres reduce the weight but also increase the strength.

Thermal and dielectric properties

Cenospheres are extremely resistant to heat and typically have a melting point in excess of 1300°C. As cenospheres are hollow, they have a relatively low coefficient of thermal conductivity, typically 0·09 W/mK. Cenospheres conduct very little electricity and are ideal for use in insulators. Specialist technical advice is available.

Physical properties

Particle size

In their raw form cenospheres have a continuous particle size distribution from submicrometre to around 500 μm. Cenospheres can be classified into particular particle size distributions by prior arrangement.

Density

The overall particle density is around 0·4–0·7 g/cm^3 but varies depending on the size fraction. Densities as low as 0·3 g/cm^3 can be achieved by selection of the various size fractions. In general, the smaller the size fraction the lower the density. Bulk density ranges from 0·25 to 0·35 g/cm^3. The density of the shell material varies between 2·0 and 2·4 g/cm^3.

Chemical composition

The cenosphere shell contains typically:

- 55–65% of SiO_2
- 25–35% of Al_2O_3
- 1–5% of Fe_2O_3
- The central voids typically contains 70% CO_2 and 30% N_2 in gaseous form.

Water absorption

Cenospheres have virtually no water-absorbing properties.

Chemical resistance

The aluminosilicate shell is particularly resistant to acidic environments. However, like most glasses, cenospheres are not resistant to strong alkalis.

Aerated ceramic

Experimental work in both the UK and the USA[12] has shown that by the addition of a foaming agent and small proportions of other materials, fly ash can be used to manufacture an aerated ceramic. Units up to $1{\cdot}2 \times 0{\cdot}6 \times 0{\cdot}1$ m have been produced in the USA.

Summary

This book has demonstrated the wide range of uses to which fly ash has been put. There are many other potential applications for this versatile and readily available by-product. Research continues throughout the world, with some 300 papers per annum being published covering the properties and applications of fly ash. It is hoped that eventually this material will be considered not a waste product but a valuable material in its own right and will be fully utilised world-wide.

References

1. Anderson M, Jackson, G. The history of pulverised fuel ash in brickmaking in Britain. *Transaction and Journal of the Insitute of Ceramics* 1987; **86**(4): 99–135.
2. BS 3921. *Specification for clay bricks*. BSI, London, 1985.
3. BS 5628. *Part 3: Code of practice for use of masonry. Materials and components, design and workmanship*. BSI, London, 1985.
4. Watts AJC. Full scale experiments on the addition of fly ash in brick making. *Transactions of the British Ceramic Society* 1954; **53**(5): 315.
5. Butterworth B. Bricks made with fly ash. *Transactions of the British Ceramic Society* 1954; **53**(5): 293.
6. British patent 653,070.
7. Cockrell CF, Shafer HE Jr, Humphreys KK. *Fly ash based structural materials: recent developments utilising the WVU-OCR process*. Report No. 16, Coal Research Bureau, West Virginia University, February 1966.
8. Jones GT, Corrie DA. *The production of 100% fly ash bricks, interim report, June 1960 to July 1961*. CEGB North Western Region Research Note No. 4/61, 1961.
9. Crimmin WRC, Gill GM, Jones GT. *The production of 100% fly ash bricks, interim report*. CEGB North Western Region Report No. 6/63, 17 June 1963.
10. CEGB. *North Western Region, fly ash news review* No. 2. CEGB, London, February 1963.
11. Cabrera JG, Zoorob SE. Design of low energy hot rolled asphalt. *Proceedings of the 1st European Symposium*, Leeds, 1994: 289–308.
12. Griffith JS, Dusek JT, Bailey EC. *A new use for fly ash – a lightweight ceramic building material*. ASME Paper No. 61-WA-291, 1961.

Index